Undergraduate Lecture Notes in Physics

Undergraduate Lecture Notes in Physics (ULNP) publishes authoritative texts covering topics throughout pure and applied physics. Each title in the series is suitable as a basis for undergraduate instruction, typically containing practice problems, worked examples, chapter summaries, and suggestions for further reading.

ULNP titles must provide at least one of the following:

- An exceptionally clear and concise treatment of a standard undergraduate subject.
- A solid undergraduate-level introduction to a graduate, advanced, or non-standard subject.
- A novel perspective or an unusual approach to teaching a subject.

ULNP especially encourages new, original, and idiosyncratic approaches to physics teaching at the undergraduate level.

The purpose of ULNP is to provide intriguing, absorbing books that will continue to be the reader's preferred reference throughout their academic career.

Series editors

Neil Ashby
University of Colorado, Boulder, CO, USA

William Brantley
Department of Physics, Furman University, Greenville, SC, USA

Matthew Deady
Physics Program, Bard College, Annandale-on-Hudson, NY, USA

Michael Fowler
Department of Physics, University of Virginia, Charlottesville, VA, USA

Morten Hjorth-Jensen
Department of Physics, University of Oslo, Oslo, Norway

Michael Inglis
Department of Physical Sciences, SUNY Suffolk County Community College, Selden, NY, USA

More information about this series at http://www.springer.com/series/8917

Gerhard Ecker

Particles, Fields, Quanta

From Quantum Mechanics to the
Standard Model of Particle Physics

 Springer

Gerhard Ecker
Fakultät für Physik
Universität Wien
Vienna, Austria

ISSN 2192-4791 ISSN 2192-4805 (electronic)
Undergraduate Lecture Notes in Physics
ISBN 978-3-030-14478-4 ISBN 978-3-030-14479-1 (eBook)
https://doi.org/10.1007/978-3-030-14479-1

Library of Congress Control Number: 2019933843

Planning/editing: Lisa Edelhäuser

This Springer imprint is published by the registered company Springer Nature Switzerland AG
The registered company address is: Gewerbestrasse 11, 6330 Cham, Switzerland

To Linda and Julia

Foreword

Why Physics?

Books often promise to be easily digestible, to provide useful counseling and exciting reading. I find this book very exciting but it requires – especially in later chapters – active engagement and willingness to learn. In this case, the reader is rewarded with a profound introduction to the foundations of modern physics.

But what is it good for? Does one really have to know about gauge groups and symmetry breaking or how quarks and gluons interact? Not necessarily. But whoever gets involved in such questions will not just get recipes for kitchen and garden but deep insights into the mysteries of nature. And once you get the taste of it you want to know more.

As a high school student in Vienna, my Christmas list contained books with such promising titles as: Einstein and the Universe, Physics and Philosophy, and Man and the Cosmos. Famous physicists like Einstein, Bohr and Heisenberg expressed themselves in their popular writings in such an easily comprehensible way that I believed to understand how modern science works.

But then I put a book with the fascinating title "Gravity and the Universe" on my reading list. The author was a certain Pascual Jordan, known to me as a pioneer of quantum theory. The bookseller handed over the slim volume with an ironic smile that I at first misunderstood as admiration. But even the first glimpse into the book shattered my naive concept of physics completely. The text mainly consisted of formulas that did not mean anything to me: Greek letters surrounded by a cloud of upper and lower indices, characters with one or two dots on top, strange symbols for – as I learned only much later – summation and integration. And as if the author wanted to ridicule me, the few lines of normal text did not explain anything but maintained: the well-known relation … holds; as can easily be seen, this follows from …, etc.

The shock made a deep impression on me, and I realised: just as one has to know Chinese in order to understand China one does not get anywhere in physics without higher mathematics. And so I decided to learn the language of nature. In other words, I decided to study theoretical physics.

Maybe that would never have happened if I had not had the physics teacher mentioned by my classmate Gerhard Ecker, author of the present book, in the introductory remarks. Of course, that excellent pedagogue had to stay within the confines of high school mathematics, but he managed, for instance, to convince us of the power of calculus in the discussion of acoustic and electromagnetic waves. We understood immediately why the cosine is the first derivative of the sine and how this can be used for the description of oscillating circuits with solenoids and capacitors.

And so Ecker and I began to study theoretical physics at the University of Vienna. At that time – I am talking about an ancient epoch in the middle of last century – the Cold War was on. The Soviet Union had just baffled the West with the start of a beeping satellite and with manned space flights. Under the impression of the Sputnik shock, the West undertook great efforts not only to catch up in aerospace but also to push forward basic science. Therefore, a career in physics was considered highly promising. However, towards the end of our studies, the general enthusiasm had started to diminish, since the U.S. had taken a clear leadership in the space race with several manned missions to the moon.

Meanwhile, I had made an experience that may be similar to that of a musician or of a chess player who realises that there are limits to his talent: sufficient for a good performance but not for championship. Such was the case with my mathematical capabilities. Therefore, after finishing my physics studies I turned to literature and science journalism, whereas my mathematically more gifted fellow student Gerhard Ecker set out to make a career in theoretical particle physics.

Thus, we embarked on separate ways, leading into the "two cultures" that the British writer and diplomat C. P. Snow had diagnosed in about the middle of last century: one characterised by literature and the humanities, the other by technology and science. Snow deplored the gap between the two cultures. In particular, he criticised that an intellectual would be despised if he did not know the trendy writers but that he did not have to know what entropy is. The mathematician Edmund Hlawka from the University of Vienna once had formulated a similar thought in an introductory lecture on calculus: "If people declare that they love Shakespeare and one of us answers that he prefers Pilsen beer he is considered an ignorant, but nobody cares what the derivative of a function is".

Snow's proposition of the two cultures has been disputed but it corresponds to my experiences as a "borderline case". My work as a science journalist consists mainly in moving back and forth between the two cultures and providing translations between the mathematically expressed findings of physics and the everyday language of literature and philosophy.

Snow formulated his hypothesis against the historical background of the Cold War and the Sputnik shock. His concern was actually a warning: if Europe would continue to favour the traditional classical-humanist ideals disregarding science, it would soon fall back behind the Soviet Union and the U.S., which produced mathematicians, scientists and engineers in great numbers. Nowadays, this concern is all water under the bridge. Today the gap between the cultures has become much smaller, and sometimes one talks about a "third culture" as a bridge. But in my view, the gap is not completely closed.

With the present book, Gerhard Ecker attempts to convey to the interested reader, from whatever culture he may come, ideas and results of modern physics in the best possible way. This involves some formulas but it only requires high school mathematics in addition to a good portion of curiosity. In the second half, the author goes considerably beyond what other books have to offer on quantum mysteries, Schrödinger's cat and spooky actions at a distance. He also goes beyond what he and I had heard in lectures during our studies. At that time, quarks were still hypothetical objects or "purely mathematical entities" as their inventor would call them. Today quarks (and gluons) are constituents of the modern theory of strong interactions, and the author uses his experience with quantum chromodynamics to introduce the reader to present-day particle physics.

I very much envy him for an experience of success that most theoretical physicists rarely make in their lifetime. In 1987, he and his collaborators used quantum field theoretic methods to predict details of an exotic particle decay that were later confirmed in accelerator experiments. Such achievements encourage the theoreticians that their work is not simply an intellectual game but that it approaches a reality that has been unfolding since the Big Bang. The great successes and the many open questions this book describes illustrate how immensely complicated nature is and nevertheless how much we already understand of it.

Aachen, Germany
September 2018

Michael Springer

Preface for the English Edition

The book presented here is essentially the English translation of my book "Teilchen, Felder, Quanten" published by Springer-Verlag in July 2017. A few small updates take into account recent developments in particle physics. In addition, glossary, index and especially the bibliography have been enlarged compared to the German version.

In the public perception of particle physics, the giant accelerator LHC at CERN dominates. In comparison, the theoretical foundation often remains in the dark and this of course is mainly due to the missing background. This book is an attempt to enable the high school graduate, the physics student in the first semesters, the physics teacher and in general all those interested in physics to retrace the development of fundamental physics during the past 120 years. Here, "fundamental physics" stands for the physics of the smallest structures of matter and their interactions.

Some readers of the German version have criticised my claim that this book can be understood with a knowledge of basic mathematics at the level of the upper secondary syllabus. While I still maintain my claim, I am of course aware that even many people interested in science will have forgotten some or most of their high school math depending on how many years have passed since graduation. To those putting the book aside after encountering the first formula, I can only ask to give it a second try after reading Appendix A. In addition, I encourage my readers to read the enlightening foreword by Michael Springer for a start.

The beginning of the acknowledgements is dedicated to the memory of my teachers Richard Lederer and Walter Thirring who shaped my approach to physics more than I was probably aware during my school and university years. During more than 50 years of my active preoccupation with physics, many more teachers, colleagues and students have accompanied and influenced me, by far too many to be listed here. I thank my family, relatives and friends for their encouragements to write this book. Among the latter is Michael Springer, schoolmate and fellow student, who accompanied the genesis of this book from the beginning to the end and who wrote an inspiring foreword. Special thanks are due to Robin Golser and

Martin Fally, Deans of the Faculty of Physics of the University of Vienna, for allowing me to use the infrastructure of the faculty even after retirement. Last but not least, I thank Stefanie Adam, Lisa Edelhäuser and Lothar Seidler from Springer-Verlag for their efficient support.

Vienna, Austria   Gerhard Ecker
September 2018

Contents

Introduction

Victor Weisskopf, one of the grand-masters of the communication of modern physics, sometimes criticised that many presentations tend to concentrate on the most recent developments, which does not necessarily lead to a deeper insight. In this spirit, superstring theory or quantum gravity will hardly play a role in this book. Instead, the development of modern physics from the beginning of the quantum era around 1900 to the radical break with classical physics through quantum mechanics, its unification with the special theory of relativity to quantum field theory up to the Standard Model of particle physics, the most comprehensive theory of physics to date, will be outlined. Following Albert Einstein's motto that one should treat matters as simply as possible but not simpler, we will have to introduce some basic mathematical formalism. The statement of Ernst Mach towards the end of the 19th century that a man without at least a rudimentary education in mathematics and science is only a stranger in this world, is politically completely inappropriate today, but it may serve as an incentive nevertheless.

Although quantum field theory, the theoretical tool of the particle physicist, is built on quantum mechanics, neither in scientific papers nor in popular accounts of particle physics will one easily find a hint as to which interpretation of quantum mechanics the particle physicist in question prefers. Whether he adheres to the Copenhagen interpretation or to the many-worlds theory or to any other variant, simply does not matter for his everyday work. Recently, the Italian particle physicist Stefano Forte has considered this issue in some detail. His treatise (Forte 2014) can especially be recommended to the philosophically inclined but also to the simple "quantum engineer" (© John Bell).

Instead of profound philosophical considerations we are going to examine a concrete example from the daily work of a particle physicist. For this purpose, the experimental and theoretical investigation of a particle decay is especially suitable. In contrast to an (elastic) scattering process, there is neither a classical limit for a

© Springer Nature Switzerland AG 2019

G. Ecker, *Particles, Fields, Quanta*, Undergraduate Lecture Notes in Physics,
https://doi.org/10.1007/978-3-030-14479-1_1

decay,[1] nor can a decay be treated in the framework of quantum mechanics (Chap. 4).
The theoretical analysis of a decay requires quantum field theoretical methods.

In general, an unstable particle has several decay channels, i.e. different final states
to which it can decay. In addition to the lifetime of the particle, the relative probability
of the decay into a definite final state is of interest. This dimensionless quantity is
called the branching ratio of the specific decay. For a two-body decay when there
are exactly two particles in the final state, this is actually the only quantity of interest
that can be calculated and measured. More interesting for the following discussion
is a many-particle decay. Therefore, we consider the decay of the neutral, long-lived
K meson into a neutral pion and two photons: $K_L^0 \to \pi^0 \gamma\gamma$. Here, long-lived is a
relative notion as the lifetime of K_L^0 is only about $5 \cdot 10^{-8}$ s, but the K_S^0 is even
shorter-lived. The branching ratio for the decay $K_L^0 \to \pi^0 \gamma\gamma$ is about $1.3 \cdot 10^{-6}$,
i.e., on average only one out of a million K_L^0 decays in this way. More interesting
for our purposes is another quantity, the invariant mass of the two photons in the
final state. This invariant mass can be calculated from the energies and momenta of
the two photons[2] and it has the big advantage that it is independent of the reference
system and therefore Lorentz invariant (Chap. 2, Appendix B). In other words, we do
not have to specify how fast the original K_L^0 mesons were before decaying. This is of
course a considerable advantage when comparing the results of different experiments
among themselves and with theoretical predictions.

The experimenters display their results in so-called histograms as shown in Fig. 1.1
for the decay in question. Since in a given experiment only a finite number of decays
can be observed, it is useful to divide the kinematically allowed range for the invariant
mass (denoted m_{34} in the specific case) on the abscissa into discrete intervals and
to investigate how many decays occur in a specific interval. The number of these
decays is then indicated on the vertical axis. In this way one obtains the typical box
structure of histograms as in Fig. 1.1. These boxes are called bins by the physicists.
It is up to the experimenter how many bins (s)he uses. However, the total number of
decays measured suggests an optimal number of bins.[3]

A fundamental aspect of quantum physics can now be examined in this decay.
If the experimenters on this side of the Atlantic (CERN, Geneva) and on the other
side (Fermilab, Batavia near Chicago) had compared – which they almost certainly
did not do –, into which bin their first decay event fell, they very probably would
have found (taking the rather big number of bins into account) that these two events
correspond to two different bins. The decay of a particle is a random event. Therefore,
this difference for the respective first decays does not imply that there is a European
and an American version of the Standard Model leading to different results. If in the
quantum field theoretical description of this decay the infamous "hidden parameters"
were missing, a "complete" description would have to be both time- and location

---

[1] In the classical limit the particle in question simply does not decay.
[2] If E_i, $\vec{p}_i$ ($i = 1, 2$) are the energies and momenta of the two photons, the invariant mass of the two
photons is defined as follows: $m_{inv} = \sqrt{(E_1 + E_2)^2 - (\vec{p}_1 + \vec{p}_2)^2 c^2}/c^2$ (c is the speed of light).
In Fig. 1.1 the invariant mass is denoted as m_{34}.
[3] In the two histograms in Fig. 1.1 the number of bins is obviously different.

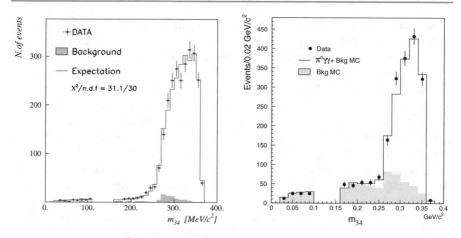

Fig. 1.1 Decay distribution in the invariant mass m_{34} of the two photons in the decay $K_L^0 \to \pi^0\gamma\gamma$. Left plot: histogram of the NA48-Collaboration (CERN) (From Lai et al. 2002; with kind permission of © Elsevier 2002. All Rights Reserved). Right plot: histogram of the KTeV-Collaboration (FNAL) (From Abouzaid et al. 2008; with kind permission of © American Physical Society 2008. All Rights Reserved). Theoretical prediction: red curve (left plot), black curve (right plot)

dependent to explain the first decays at CERN and at Fermilab. Maybe mythology could provide such a complete description, but not a universal science.

The second decay in each experiment is almost as inconclusive as the first one, but in the course of time the bins get more and more populated and the two histograms begin to resemble each other more and more. On the other hand, with the quantum field theoretical methods of the Standard Model one can calculate a (continuous) probability distribution that can be projected onto the bins chosen by the experimenters. This distribution is also displayed in the two plots of Fig. 1.1. It had actually been calculated already before the experimental verification (Ecker et al. 1987). The agreement between the theoretical prediction and the experimental results allows us to claim that we "understand" the distribution in the invariant mass of the two photons in the decay $K_L^0 \to \pi^0\gamma\gamma$ in the framework of the Standard Model, not more, but also not less.

Our decay can also serve as an illustrative example for the many-worlds interpretation of quantum mechanics (Everett 1954). For the apostles of Everett, the appearance of the first or of any decay in a certain bin only means that in other worlds this decay could occur in other kinematically allowed bins. Those other worlds are causally disconnected from our world. Therefore, no information can be exchanged between different worlds à la Everett. Most experimenters are probably unaware that in choosing his bins the physicist can be viewed as a sort of demiurge, a creator of worlds. As long as we cannot communicate with the other worlds, this interpretation can at least not do any harm. On the other hand, it tells us nothing about the structure of matter and interactions.

Advances in our knowledge of the structure of matter and of its interactions are sometimes exemplified with the help of the so-called quantum ladder, a concept often used by Victor Weisskopf. This ladder stands on the foundations of classical physics. The actual quantum ladder starts with the first rung corresponding to atomic physics and its theoretical basis quantum mechanics. From the first step upwards the quantum connection between available energy and resolution holds. As we step up the ladder, the characteristic energy increases and with it the resolving power of ever smaller structures. At a given step, i.e. with the energy corresponding to this step, we can only resolve structures at this and all lower rungs. Even if it should turn out one day that the matter particles leptons and quarks are composed of more fundamental constituents, we will not need to know this possible substructure of leptons and quarks in order to understand the hydrogen atom. The theoretical description reflects this basic fact. Nuclear physics and the structure of atomic nuclei in terms of protons and neutrons (nucleons) are located on the next-higher rung. Particle physics still one step higher enabled the resolution of the substructure of nucleons with quarks and gluons as constituents. Moreover, particle physics led to the realisation that the three fundamental interactions of the microcosm (electromagnetism, weak and strong nuclear forces) can all be described in terms of gauge theories in the framework of the Standard Model. A possible higher rung is still hidden in the clouds. Whether there is such a higher rung and what it might look like, will be discussed at the end of this book.

In Chap. 2 some of the problems of classical physics towards the end of the 19th century will be discussed. The quantum era starts in the year 1900 with Planck's radiation law. In that chapter we also review the special theory of relativity, which, although not relevant for quantum mechanics, will be essential for quantum field theory. Starting from the stability problem of Rutherford's model of the atom, Bohr presents his planetary model of the atom in 1913. As intuitive as it may appear, Bohr's atomic model is only an intermediate step on the way to quantum mechanics that we retrace in Chap. 3. The two versions of quantum mechanics put forward in 1925/26, Heisenberg's matrix mechanics and Schrödinger's wave mechanics, turned out to be equivalent. In addition to the Schrödinger equation we also "derive" the uncertainty relation and discuss its interpretation. Introduction of the spin of the electron leads to the Dirac equation, the relativistic generalisation of the Schrödinger equation. The Dirac equation suggests the existence of antimatter and it explains the fine structure of the hydrogen atom. Causality is violated in quantum mechanics, which even in the relativistic form of the Dirac equation can only be applied to processes where particles are neither created nor annihilated. We perform the crucial step for particle physics from quantum mechanics to relativistic quantum field theory in Chap. 4. Quantum field theory explains the connection between spin and statistics (bosons vs. fermions) and it allows for the derivation of the fundamental CPT theorem. The crucial role of symmetries in particle physics will also be discussed in the framework of quantum field theory.

Quantum electrodynamics (QED) is introduced in Chap. 5. It is the quantum version of Maxwell's electrodynamics describing the interaction between charged particles and the quantised electromagnetic field. Historically, QED is the first example of a gauge theory. For the comparison between theory and experiment, S-matrix elements are calculated, the quantum field theoretical analogue of the nonrelativistic wave function. The calculations are performed in the framework of perturbation theory, an expansion in powers of the fine-structure constant. We introduce the graphical representation of S-matrix elements in terms of Feynman diagrams, with application to the Compton scattering of photons on electrons. The calculation of the anomalous magnetic moment of the electron up to the fifth order is one of the triumphs of the perturbative treatment of QED. The crisis of quantum field theory, in particular of QED, in the 1930s and in the first half of the 1940s is treated in Chap. 6. This crisis was caused by the divergences (infinities) of perturbation theory occurring in most processes beyond lowest order. The solution of the crisis was based on a manifestly Lorentz invariant perturbation theory and on the concept of renormalisation. After renormalisation the unknown structure of the theory at shortest distances and highest energies is only contained in masses and coupling constants, which must therefore be determined experimentally. We also comment on the originally widespread skepticism towards the renormalisation program. In Chap. 7 we retrace the developments from the nonrenormalisable Fermi theory of the weak interaction to the unified electroweak gauge theory. Gauge symmetry and its spontaneous breaking are the key elements of this unification. Parity violation of the weak interaction is discussed in detail. The strong interaction is covered in Chap. 8. Until the beginning of the 1970s a perturbative treatment of the strong interaction seemed hopeless. Experimental indications pointing towards a weakening of the strong force at high energies and the discovery of asymptotic freedom in non-abelian gauge theories paved the way for quantum chromodynamics (QCD) that can be treated perturbatively at high energies.

The electroweak gauge theory and QCD together constitute the Standard Model of fundamental interactions treated extensively in Chap. 9. The evidence for exactly three generations of fundamental fermions (leptons and quarks) is discussed. The simplest mechanism for the spontaneous breaking of electroweak gauge symmetry seems to be realised in nature as indicated by the discovery of the Higgs boson in 2012. Although the Standard Model presently agrees with all experimental results even at the highest LHC energies, most particle physicists are convinced that the Standard Model cannot be the final theory of the fundamental interactions. The discovery of neutrino oscillations requiring at least two massive neutrino types is a first indication for an extension of the Standard Model. This phenomenon and additional arguments for an underlying structure such as grand unification are treated in Chap. 10. The grand unification of strong, electromagnetic and weak interactions would allow for a deeper understanding of the structure of the microcosm, but tangible experimental findings such as proton decay are still missing. In the final Chap. 11 we collect experimental and theoretical approaches for finding hints for "New Physics" beyond the Standard Model. For this purpose we survey some promising projects of experimental high energy physics. The Standard Model most probably needs to be modified at higher energies. It is therefore viewed as an effective quantum field theory valid at presently available energies. The concept and applications of effective field theories also at lower energies conclude this final chapter. Appendix A contains some mathematical structures used in the book. The system of units generally adopted in particle physics and some orders of magnitude are also reviewed in Appendix A. Gauge and Lorentz transformations are discussed in some detail in Appendix B. Appendix C contains keyword-type biographies of scientists cited in the text. A glossary and an index conclude the book.

References

Abouzaid E et al (KTeV-Collaboration) (2008) Phys Rev D 77:112004. arXiv:0805.0031
Ecker G, Pich A, de Rafael E (1987) Phys Lett B 189:363
Everett H (1954) Rev Mod Phys 29:454
Forte S (2014) Euresis J 6:49. arXiv:1309.1996
Lai A et al (NA48-Collaboration) (2002) Phys Lett B 536:229. arXiv:hep-ex/0205010

Physics Around 1900

2

Physics Before 1900

In the second half of the 19th century the world of classical physics was still in order. This was spelled out by the physicist Philipp von Jolly who in 1874 advised the young Max Planck against studying theoretical physics (Planck 1933): "Physics is a highly developed and nearly fully matured science …There may still be a speck of dust or a vesicle here and there that needs investigation but the system as a whole is rather well established. Theoretical physics notably approaches the degree of perfection that geometry for instance has been enjoying for centuries.[1]"

Based on the scientific insights of Galileo Galilei and Johannes Kepler, Isaac Newton performed the first great synthesis of physics in his opus magnum "Mathematical principles of physics" (1687). The fall of his famous apple from a tree and the planetary orbits obey the same universal law of gravitational attraction. Especially by Joseph-Louis Lagrange at the end of the 18th century and then by William Hamilton in the first half of the 19th century, classical mechanics was given its final form.

Newton's authority also manifests itself in the development of the theory of light. Since Newton had declared light to be composed of small particles, the competing wave theory disappeared from the scene for about 100 years. But at the beginning of the 19th century clear indications for the wave nature of light (interference, diffraction, polarisation) appeared. In parallel to those discoveries, the manifestations of electricity and magnetism were investigated. The second great synthesis of physics was carried out in 1864 by James Maxwell who followed up the concepts of electric and magnetic fields by Michael Faraday. In his theory of electromagnetism (Maxwell equations, Appendix B) he combined three main areas of physics: electricity, mag-

---

[1] Translation from the original German by the author.

© Springer Nature Switzerland AG 2019
G. Ecker, *Particles, Fields, Quanta*, Undergraduate Lecture Notes in Physics,
https://doi.org/10.1007/978-3-030-14479-1_2

netism and optics. The experimental demonstration by Heinrich Hertz that both radio waves and light are electromagnetic waves differing only in their wave lengths was instrumental for the general acceptance of Maxwell's theory.

Towards the end of the 19th century most physicists therefore shared the view that the fundamental laws of physics were known and that future generations would only have to apply those laws. Less known than the previously quoted advice of Planck's teacher but all the more surprising is the opinion of the American physicist Albert Michelson from 1899 (cited in Teich and Porter 1990): "The more important fundamental laws and facts of physical science have all been discovered, and these are so firmly established that the possibility of their ever being supplanted in consequence of new discoveries is exceedingly remote." This is all the more surprising as the experiment of Michelson and Morley from 1887 dealt a fatal blow to the ether as carrier of electromagnetic waves. Nearly 20 years later this was an experimentum crucis for Albert Einstein to put forward his special theory of relativity.

However, more and more "vesicles" if not cracks began to show up in the grand architecture of classical physics.

- In the debate on the physical reality of atoms (Ernst Mach: "Have you ever seen one?") the atomistic view, advocated especially by Ludwig Boltzmann, began to prevail.
- In spite of a lot of negative evidence, the ether as carrier of electromagnetic waves was seemingly hard to kill. Because of the experimental evidence it had to be equipped with strange properties such as the shortening of measuring rods moving relative to the ether (Lorentz contraction).
- The structure of atomic spectral lines was a complete puzzle for classical physics. Already in 1885 Johann Balmer had discovered empirically a formula for the spectral lines in the visible spectrum of the hydrogen atom, generalised three years later by Johannes Rydberg. The inverse wave lengths were found to obey the following equation:

$$\frac{1}{\lambda_{nm}} = R\left(\frac{1}{n^2} - \frac{1}{m^2}\right), \qquad n \geq 1, \, m \geq n+1 . \tag{2.1}$$

R is an empirical constant (Rydberg constant) that would find its explanation in Bohr's atomic model and then in quantum mechanics. The Balmer series corresponds to $n = 2$. The formula (2.1) was later experimentally confirmed for $n = 1$ (Lyman series, ultraviolet spectrum), $n = 3$ (Paschen series, infrared spectrum), etc. The energy levels of the hydrogen atom are displayed in Fig. 2.1.

- Already in 1892 Hendrik Lorentz had predicted the splitting of spectral lines in a homogeneous magnetic field. A few years later Pieter Zeeman confirmed this splitting experimentally, but the splitting was substantially bigger than predicted by Lorentz. Lorentz himself found the explanation. The Lorentz force on moving charges in a magnetic field does not act on the atom as a whole but instead on the electrons discovered shortly before. In comparison with the atom, the mass of the electron is much smaller and this explains the size of the Zeeman effect because

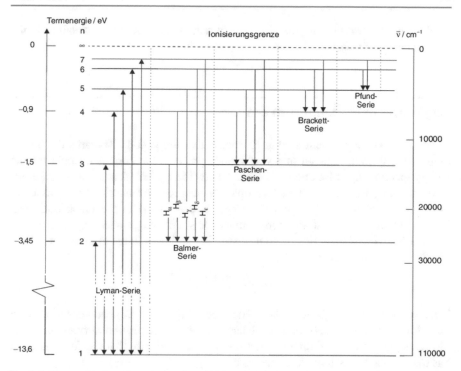

Fig. 2.1 Energy levels and spectral series of hydrogen. The energy of the levels is given in eV. The energy of the ground state with principal quantum number $n = 1$ is $-13.6\,eV$ (Eq. (3.10)) (From Demtröder 2010; with kind permission of © Springer-Verlag Berlin/Heidelberg 2010. All Rights Reserved)

the magnetic moment relevant for the splitting is inversely proportional to the mass (Eq. (5.5)). However, this only explained the so-called "normal" Zeeman effect (splitting into three lines), the much more frequent "anomalous" effect can only be understood in the framework of quantum mechanics due to the spin of the electrons. From today's point of view the terminology is a historical curiosity. The "normal" effect is a special case of the more general "anomalous" effect for total spin zero of the electrons.

- On the basis of his experiments with cathode rays, J. J. Thomson in 1897 postulated the existence of "corpuscles", which were subsequently renamed electrons by Lorentz.
- In 1896 Henri Becquerel discovered a mysterious radiation that was called radioactivity by Marie and Pierre Curie and investigated by them in detailed experiments. Soon one distinguished between α, β and γ rays. Ernest Rutherford and Frederick Soddy realised in 1902 that radioactivity has to do with the emission of particles from the atomic nucleus obeying the radioactive decay law. α rays (emission of helium nuclei) and β rays (emission of electrons) also involve a transmutation of

elements.[2] It was obvious that this type of radiation could not be explained in the framework of classical physics.

Beginning of the Quantum Era

There is general agreement among physicists that the year 1900 marks the beginning of the quantum era. In that year Planck presents the correct formula for the energy spectrum of a black body. The black body is a typical idealisation of theoreticians standing for an object that completely absorbs and maximally emits electromagnetic radiation. The spectral energy density[3] $\eta(\nu)$ in the black body can be derived from classical electrodynamics for small frequencies[4] ν. The result[5] ($k = 1.38064852(79) \cdot 10^{-23}$ J(oule)/K(elvin) is the Boltzmann constant, T the absolute temperature in K)

$$\eta(\nu) = \frac{8\pi k T}{c^3}\nu^2 \tag{2.2}$$

is known as the Rayleigh–Jeans law. The spectral density (2.2) becomes arbitrarily large for large frequencies (small wave lengths), an absurd result deserving the name ultraviolet catastrophe. Wilhelm Wien observed empirically that the formula (2.2) has to be replaced for large frequencies by

$$\eta(\nu) = a\nu^3 e^{-b\nu/T} \, , \tag{2.3}$$

with two parameters a, b to be determined experimentally. In October 1900 Planck presents an empirical interpolation formula that reads in today's notation

$$\eta(\nu) = \frac{8\pi h\nu^3}{c^3} \frac{1}{e^{h\nu/kT} - 1} \, , \tag{2.4}$$

with an a priori undetermined constant h. In December of the same year Planck submits a theoretical justification for his radiation law (Planck 1901): the exchange of energy in the black body can only happen in packages of size $E = h\nu$. Since then the constant $h = 6.626070040(81) \cdot 10^{-34}$ Js is known as Planck's constant. For Planck only the energy transfer but not the radiation itself was quantised.

---

[2]Soddy: "Rutherford, this is transmutation!"; Rutherford: "Don't call it transmutation. They'll have our heads off as alchemists." (Cited in Howorth 1958).

[3]Integrating the spectral density over all frequencies, one obtains the total energy density of the black body.

[4]As a reminder, frequency ν and wave length λ satisfy the relation $\lambda\nu = c$ where c is the speed of light.

[5]Physical units and the notation for numerical values are explained in Appendix A. Unless specified otherwise, all numerical values in this book are taken from the Review of Particle Properties (Tanabashi et al. 2018).

An actual derivation of the radiation formula (2.4) is only possible with the help of quantum statistics. For small arguments x the exponential function can be expanded as $e^x = 1 + x + \cdots$. Thus, for small frequencies $e^{h\nu/kT} - 1 = h\nu/kT + \cdots$ and in this limit Planck's formula (2.4) turns into the Rayleigh–Jeans law (2.2). This derivation also shows that (2.2) is the classical limit ($h \to 0$) of (2.4). The derivation of Wien's law (2.3) is even simpler if one recalls that for large frequencies $e^{h\nu/kT} \gg 1$.

To get a feeling for the orders of magnitude involved, we consider small oscillations of a pendulum with mass $0.1\,\text{kg}$ and length $1\,\text{m}$. The product of the average energy $\langle E \rangle$ of the pendulum and the time duration of an oscillation τ is then given by $\langle E \rangle \tau \sim 10^{34}\, h\, \varphi_0^2$ where φ_0 is the maximal deflection angle. The action of the pendulum is therefore larger than Planck's constant h by so many orders of magnitude that one can safely neglect all quantum effects. At the same time we observe that energy $\times$ time and incidentally also length $\times$ momentum have the dimension of an action.

In 1905, the annus mirabilis of physics, Einstein publishes four ground-breaking articles in Annalen der Physik. The first of those papers provides an interpretation of the photo effect (Einstein 1905a) with the hypothesis of light quanta.[6] In the photo effect electrons are extracted from the surface of metals by shining short-wave light on the surface. The experimental results disagreed with the classical wave theory of light. For instance, the kinetic energy of the emerging electrons does not depend on the intensity of the radiation but only on the frequency of the light. The minimal frequency for extracting electrons depends on the material of the metal surface. Einstein went beyond the ideas of Planck and postulated that light itself consists of discrete quanta with energy $E = h\nu$, which release the electrons in elementary processes. With this hypothesis all phenomena of the photo effect could be explained. Moreover, since there also exists unambiguous evidence for the wave nature of light we are confronted for the first time with wave-particle duality. The photo effect is therefore a key experiment for the foundation of quantum physics. For his hypothesis of light quanta, Einstein received the Nobel Prize of 1921.

To illustrate the relevant orders of magnitude, let us consider light from the sun. The sun light impinging on the earth with a clear sky corresponds to about $4 \cdot 10^{21}$ photons/m² s (i.e. per square meter of the earth surface and per second). The human eye, on the other hand, reacts already to much smaller photon fluxes. The evolution ensured that our eyes are especially sensitive to the part of the electromagnetic spectrum corresponding to sun light. The maximal sensitivity occurs for a wave length of $555\,\text{nm}$. At this wave length a healthy eye adapted to darkness reacts to a flux of about 10 photons/s. If we had not known anyway, the comparison with the solar photon flux would have convinced us not to look at the sun with unprotected eyes.

Already in 1827 the Scottish botanist Robert Brown discovered what came to be called Brownian motion, the irregular movements of small particles in fluids visible

---

[6]Instead of light quanta, Arthur Compton propagated the name photons proposed by the physical chemist Gilbert Lewis in 1926.

in the microscope. Even before 1905 it was clear that this phenomenon was caused by irregular thermal movements of even smaller particles colliding continuously with the particles observed in the microscope. In his thesis Einstein (1905b) did not remain at this qualitative level. On the basis of the molecular theory of heat, he made the quantitative prediction that on average the colliding particle covers a certain distance whose square is proportional to the observation time and to the temperature and inversely proportional to the radius of the particle and to the viscosity of the fluid. This prediction was experimentally confirmed by Jean-Baptiste Perrin during the following years. This also closed the case of the reality of atoms and the size of the invisible molecules could be estimated to $\gtrsim 10^{-10}$ m.

Special Theory of Relativity

Finally, in the last two papers (Einstein 1905c, d) of his annus mirabilis Einstein formulates the special theory of relativity (SR) that fundamentally changed our perceptions of space and time. Although it did not have any direct influence on the development of quantum mechanics, which is a nonrelativistic theory, it was of eminent importance for quantum field theory. All experimental observations of the last 110 years show that our world is relativistic. Consequently, our theories of the fundamental interactions must be compatible with SR. In the physicist's terminology those theories must be Lorentz invariant: the underlying (field) equations remain unchanged under Lorentz transformations (Appendix B). The corresponding principle of relativity also exists in classical mechanics. All inertial systems are on equal footing and therefore the Newtonian equations of motion have the same form in all inertial systems. These distinguished reference systems are defined in Newton's first axiom: there exist reference systems with a universal time (inertial systems) where in the absence of forces all point particles have constant velocities (uniform motion along straight lines). Once again, the alleged existence of inertial systems is a theoretical idealisation. The following reference systems are increasingly better approximations to an inertial system: earth, space ship, solar system, system of fixed stars, ...In classical mechanics different inertial systems are related by a class of transformations (Galilei transformations (B.12)) with the postulated universal or absolute time. Passing from one inertial system to another differing by a relative constant velocity $\vec{v}$, all velocities in Newtonian mechanics change by precisely this velocity $\vec{v}$ (law of velocity addition (B.14)). Einstein cuts the Gordian knot of the ether problem and postulates that the speed of light has the same value c in all inertial systems. But this is incompatible with Galilean invariance because by a proper choice of the velocity $\vec{v}$ in the Galilei transformation one could obtain an arbitrarily large speed of light. But then also the notion of an absolute time is no more tenable and the Galilei transformation must be replaced by the Lorentz transformation (B.11). The dependence of time on the reference system is at the root of almost all conceptual difficulties of SR such as the twin paradox (see below).

Lorentz transformations were actually known before Einstein. Henri Poincaré who also suggested the name gave Lorentz transformations their final form shortly before the publication of SR (Poincaré 1905). In the same year Poincaré also showed that Maxwell's electrodynamics is Lorentz invariant, i.e., the Maxwell equations (B.1) have the same form in all inertial frames. However, both Lorentz and Poincaré adhered to the existence of an ether. For them Lorentz transformations defined the transitions from the special reference system with a stationary ether to all other reference systems. Thus, physicists before Einstein distinguished between true and apparent coordinates. The equivalence of inertial systems in SR made the ominous ether not only unobservable but also irrelevant. Nevertheless one can imagine why Einstein's achievements were often put into question, sometimes in connection with antisemitic undertones.

In his second paper on SR (Einstein 1905d), Einstein investigated the consequences of his requirement that all physical theories must be Lorentz invariant. Maxwell's equations for electrodynamics satisfy this condition automatically (it is not clear whether Einstein was aware of Poincaré's proof), but Newtonian mechanics must be modified. One consequence of this requirement is the most famous formula of physics:

$$E = mc^2 . \tag{2.5}$$

This relation between the energy and the mass of a particle has often been a source of confusion. To emphasise that a particle at rest has an energy mc^2, this equation is sometimes written in the form $E_0 = mc^2$. For historical reasons, in many school books and even in some text books the mass m is unfortunately replaced by a so-called rest mass m_0. In spite of its alleged clarity this rest mass and the related "dynamical" mass are confusing and misleading.[7] Energy and momentum of a particle depend on the chosen inertial frame just like temporal and spatial coordinates because they are modified by Lorentz transformations. The mass, on the other hand, is the same in all inertial systems, it is a Lorentz invariant quantity. This does not imply that mass cannot change. As nuclear fission, fusion or the annihilation of matter with antimatter demonstrate, mass can be transformed into energy and vice versa, but such transformations have nothing to do with different inertial systems. They require the framework of nuclear or particle physics for a satisfactory treatment.

Before setting off again for the road to quantum mechanics, we briefly deal with the conceptual difficulties of SR, which mostly have to do with the departure from Newton's absolute time. The claim of SR that time progresses more slowly in moving systems than in the rest system of the observer (time dilatation) is admittedly difficult to swallow. For velocities small in comparison with the speed of light, Lorentz transformations resemble Galilei transformations. An equivalent formulation, maybe only for physicists, is that in the limit $c \to \infty$ a Lorentz transformation turns into a Galilei transformation (Appendix B). In any event it is clear that the most impressive manifestations of time dilatation will occur for velocities in the vicinity of the speed

---

[7]A detailed discussion can be found in Okun (1989).

of light. This is generically the case in particle physics, which therefore provides the best examples for time dilatation.

Muons belong to the class of leptons and can be viewed as more massive siblings of electrons. They are unstable and they decay (in their rest system) with an average lifetime of $\tau_\mu \simeq 2.2 \cdot 10^{-6}$ s. They are generated in the atmosphere by cosmic rays and as massive particles they are always slower than the velocity of light. Therefore, one would expect them to decay after a distance of at most $c\tau_\mu \simeq 660$ m. More precisely, their number should be reduced to the eth part with Euler's number $e \simeq 2.71828$. In reality, muons reach the surface of the earth even if they are produced at altitudes as high as 30 km. Seen from the earth, muonic "clocks" produced in the atmosphere seem to be slower and, in fact, considerably slower because the muons travel almost with the speed of light. Time has no longer any absolute meaning, it depends on the reference system.

Even more spectacular is the following biologically inoffensive experiment on the twin paradox. To measure the magnetic moment of muons, they were kept on a circular trajectory by a magnetic field in the storage ring of the Brookhaven National Laboratory (Bennett et al. 2006). Again one would naively expect that on average the muons would have decayed after a distance of around 660 m, which corresponds to about 7.5 circles in the Brookhaven storage ring. However, since they travel with almost the velocity of light ($v \simeq 0.99942\,c$) SR predicts that in the local time of the experimenter they should only decay (always according to the radioactive decay law) after

$$T = \frac{\tau_\mu}{\sqrt{1 - v^2/c^2}} \simeq 29.3\,\tau_\mu \; . \tag{2.6}$$

That corresponds to almost exactly 220 rounds in the storage ring as actually observed. Incidentally, this is also a very precise measurement of the time dilatation factor $1/\sqrt{1 - v^2/c^2}$. What has all of this to do with the twin paradox? If next to the storage ring the experimenter sets up a normal muon decay experiment with muons essentially at rest, (s)he will observe that the muons in the storage ring live 29.3 times longer than their stationary siblings. Here the frequent discussion whether after his return the traveling twin is "really" younger than his sister at home is out of place. The difference in the lifetimes of the muons is an undeniable experimental fact.

Rutherford's Model of the Atom

In the first decade of the 20th century the atom was pictured as a (positively charged) dough with the (negatively charged) electrons embedded like raisins (Thomson's plum pudding model). In the years 1908–1913, Johannes Geiger and Ernest Marsden, following suggestions by Rutherford, performed a series of scattering experiments with α particles (helium nuclei) impinging on thin foils of gold and other metals. The main results were the following:

- Most α particles passed the foils freely.
- The bigger the scattering angle, the fewer particles were scattered.
- One of approximately 8000 α particles was completely back-scattered.

On the basis of these results, Rutherford concluded in 1911 that nearly all the mass of the atom must be concentrated in the positively charged nucleus with radius $\gtrsim 10^{-15}$ m. The much lighter negatively charged electrons circle around the nucleus like planets around the sun to make the atom overall electrically neutral (Rutherford 1911). But since atoms are about 100000 times bigger than their nuclei they must consist predominantly of empty space. In the following years, the physicist's perception of an atom was dominated by Rutherford's atomic model. However, the model could not explain the structure of spectral lines nor could it answer the question why the electrons do not fall into the nucleus.

Following a suggestion by Max von Laue, Friedrich et al. showed that X-rays were diffracted on crystal lattices (Friedrich et al. 1912), a clear indication for the wave nature of radiation. Moreover, the diffraction pattern confirmed both the regular structure of matter and the size of atoms with $r_{\text{Atom}} \gtrsim 10^{-10}$ m $= 1$ Å(ngström).

References

Bennett GW et al (Muon $g-2$ Collaboration) (2006) Phys Rev D73, 072003. arXiv:hep-ex/0602035

Demtröder W (2010) Experimentalphysik 3: Atome, Moleküle und Festkörper. Springer Spektrum, Berlin

Einstein A (1905a) Ann Phys 17:132

Einstein A (1905b) Ann Phys 17:549

Einstein A (1905c) Ann Phys 17:891

Einstein A (1905d) Ann Phys 18:639

Friedrich W, Knipping P, Laue M (1912) Sitzungsberichte K Bayer Akad Wiss, math-phys Klasse 303

Howorth M (1958) Pioneer research on the atom: the life story of Frederick Soddy. New World, London

Okun LB (1989) The concept of mass. Phys Today 42:31

Planck M (1901) Ann Phys 4:553

Planck M (1933) Wege zur physikalischen Erkenntnis: Reden und Vorträge. Verlag S Hirzel, Leipzig

Poincaré H (1905) C R 140:1504

Rutherford E (1911) Phil Mag 21:669

Tanabashi M et al (Particle Data Group) (2018) Phys Rev D 98:030001

Teich M, Porter R (1990) Fin de Siècle and its legacy. Cambridge University Press, Cambridge

The Path to Quantum Mechanics

3

Bohr's Atomic Model

Niels Bohr finished his studies at the University of Copenhagen in 1911 with a thesis on the magnetic properties of metals. In September of the same year he went to Cambridge to continue his studies at the famous Cavendish Laboratories with Thomson. Thomson received him cordially and seemed to be interested in the work of the young Dane. However, in the literature one can find the cryptic remark that their communication was hampered by language barriers; maybe Thomson's knowledge of Danish was only rudimentary ...While Bohr still had to grapple with these difficulties, Rutherford visited Cambridge and reported on his new insights concerning the structure of atoms. Bohr was fascinated by Rutherford and decided to move to Manchester. To the surprise of many colleagues soon also Rutherford was quite impressed by Bohr. This was not necessarily to be expected because in general Rutherford did not have a very favourable opinion of "pure" theoreticians. After all, he had not needed a theoretician to deduce the structure of atoms from the scattering experiments in Manchester. When asked why he made an exception for Bohr, he is reported to have said: "Bohr is different, he is a football player."

During the four months of his stay in Manchester, Bohr suspected the existence of isotopes on the basis of experiments in Rutherford's laboratory, i.e. elements with the same nuclear charge but with different masses. Rutherford was not convinced and admonished Bohr not to make speculations without concrete experimental evidence. It seems that Bohr never openly complained about this discouragement, not even when Soddy received the Nobel Prize in 1921 for the discovery of isotopes. Instead he turned to Rutherford's atomic model and considered especially the problem of stability. According to classical physics, electrons should lose their energy through the emission of (synchrotron) radiation and fall into the nucleus. Bohr was convinced that the solution of this serious problem was related to the quanta of Planck and Einstein. In the spring of 1912 he set to work and published his results (Bohr's

© Springer Nature Switzerland AG 2019
G. Ecker, *Particles, Fields, Quanta*, Undergraduate Lecture Notes in Physics,
https://doi.org/10.1007/978-3-030-14479-1_3

atomic model) finally in 1913 in three articles in the Philosophical Magazine (Bohr 1913).

Besides the problem of stability, theory was also faced with the question of the size of atoms. Experiments had already answered the question ($\gtrsim 10^{-10}$ m), but theory was still lagging behind. For a first attempt one could try to combine the classical parameters e (elementary charge), m_e (mass of the electron) and c (speed of light) to construct a quantity with the dimension of a length. Up to a numerical factor, this length is unique. One possibility is to equate the absolute value of the potential energy of an electron in the hydrogen atom with its rest energy[1]:

$$\frac{e^2}{4\pi r} = m_e c^2. \tag{3.1}$$

The length resulting from this formula,

$$r_{cl} = \frac{e^2}{4\pi m_e c^2} \simeq 2.8 \cdot 10^{-15}\,\text{m}, \tag{3.2}$$

is called the classical electron radius and it has obviously nothing to do with the size of the hydrogen atom. The term classical electron radius is still used because after all Planck's constant does not appear in Eq. (3.2). However, it is highly misleading since at distances of the order 10^{-15} m classical physics is no more valid. After this failure we divide r_{cl} by the fine-structure constant α to obtain again a length that is known as the Compton wave length of the electron:

$$r_C = r_{cl}/\alpha = \frac{\hbar}{m_e c} \simeq 3.9 \cdot 10^{-13}\,\text{m}. \tag{3.3}$$

The Compton wave length will play a role in particle physics but it is still too small for an atomic radius. But dividing by α goes in the right direction and so we try once more. The result is called the Bohr radius and it is indeed the correct answer for the size of the hydrogen atom:

$$r_B = r_C/\alpha = r_{cl}/\alpha^2 = \frac{4\pi \hbar^2}{m_e e^2} = 0.52917721067(12) \cdot 10^{-10}\,\text{m}. \tag{3.4}$$

While both quantities r_{cl} and r_C contain the speed of light, r_B does not depend on c. This is a first indication that Bohr's atomic model is nonrelativistic. Despite our successful manoeuvring with powers of α the question is legitimate which physical arguments Bohr used to derive (3.4).

---

[1]We use the Heaviside system commonly employed in particle physics (see Appendix A for a detailed description) where the dimensionless fine-structure constant $\alpha \simeq 1/137.036$ has the form $\alpha = e^2/(4\pi \hbar c)$ with $\hbar = h/2\pi$.

In Bohr's atomic model of the hydrogen atom the electron moves on a circular orbit around the positively charged nucleus. Therefore, the Coulomb force is equal to the centrifugal force:

$$\frac{e^2}{4\pi r^2} = \frac{m_e v^2}{r} , \tag{3.5}$$

where v is the velocity of the electron on its orbit. The angular momentum – more precisely, its absolute value – is then equal to

$$L = m_e v r \tag{3.6}$$

and it has the dimension of an action! Although unlike Planck and Einstein we are not concerned with electromagnetic radiation here, Bohr postulated that also the angular momentum of the electron is quantised:

$$L = n\hbar \quad (n = 1, 2, \ldots) . \tag{3.7}$$

From (3.5), (3.6) and (3.7) we derive

$$r = \frac{4\pi n^2 \hbar^2}{m_e e^2} \longrightarrow r(n = 1) = r_B = \frac{4\pi \hbar^2}{m_e e^2} , \tag{3.8}$$

and the radius of the innermost circle (ground state of the hydrogen atom) is given indeed by the Bohr radius r_B. Thus, the electron does not come closer to the nucleus than the Bohr radius and the stability of the H-atom seems to be saved. From the same equations the velocity of the electron on the nth orbit turns out to be

$$v = \frac{n\hbar}{m_e r} = \frac{e^2}{4\pi \hbar n} = \frac{\alpha c}{n} \ll c , \tag{3.9}$$

justifying a posteriori the nonrelativistic treatment. Remembering that the total energy is the sum of potential and kinetic energy, one obtains for the energy of the electron on the nth orbit

$$E_n = \frac{m_e v^2}{2} - \frac{e^2}{4\pi r} = -\frac{1}{n^2} \frac{\alpha^2 m_e c^2}{2} = -\frac{E_R}{n^2} \simeq -\frac{1}{n^2} 13.6\,\text{eV} . \tag{3.10}$$

This value for the energy is substantially smaller than the rest energy of the electron, confirming once more the legitimacy of the nonrelativistic treatment. Writing the ground state energy E_R in the form $E_R = h c R$, for the transition from the mth to the nth level ($m > n$) one obtains a radiation frequency

$$\nu_{nm} = (E_m - E_n)/h = c R \left(\frac{1}{n^2} - \frac{1}{m^2} \right) , \tag{3.11}$$

corresponding exactly to Rydberg's formula (2.1). The numerical value of the Rydberg constant

$$R = \frac{\alpha^2 m_e c}{2h} \simeq 1.1 \cdot 10^7 / \text{m} \tag{3.12}$$

agrees with the measured value up to small corrections.

Bohr's atomic model immediately had resounding success. Although very soon difficulties showed up, this "old" quantum theory is still popular today because it is so nicely compatible with our classical picture of electrons moving like planets around the nucleus in the center. In the following 12 years many physicists, among them especially Arnold Sommerfeld (hence often Bohr–Sommerfeld model), tried to remove the deficiencies of the model. In the end, all those attempts were in vain and the frustration grew. A nice quote describing the general feeling is a remark of Max Born in a letter to Einstein in October 1921 (Einstein et al. 1969): "Die Quanten sind eine hoffnungslose Schweinerei.[2]"

Which were the main problems of the Bohr–Sommerfeld model?

i. The model successfully described atoms or ions with a single electron but it failed already for the helium atom (two electrons).
ii. The quantisation of angular momentum was confirmed by quantum mechanics but the angular momentum L was too big by the value of $\hbar$ for all stationary states of the H-atom. In particular, in the ground state $(n = 1)$ $L = 0$ (and not $L = \hbar$ as predicted by Eq. (3.7)).
iii. The idea of well-defined electron orbits all in one plane (disk model) is not consistent with quantum mechanics, which predicts a finite probability density for the electron everywhere in the atom. Moreover, the planetary picture of Bohr's model cannot explain chemical binding.
iv. The anomalous Zeeman effect (splitting of spectral lines in a magnetic field) could not be explained.

In the years 1912–1914 James Franck and Gustav Hertz carried out experiments confirming the existence of discrete energy levels (Franck and Hertz 1914). Electrons accelerated by an electric field collide with atoms, thereby losing energy. The dependence on the acceleration voltage and investigations of the light emitted by the excited atoms showed that the energy transfer, both emission and absorption, proceeds in discrete packages, a spectacular support for Bohr's atomic model and for quantum theory in general.

About ten years later (1922–1923), Compton confirmed the existence of photons in experiments scattering light on matter (Compton 1923, Fig. 3.1). Monochromatic X-rays (i.e. with well-defined wave length) are scattered on crystals. Differently from classical electrodynamics, the scattered radiation possesses a greater wave length than the original radiation. The difference of wave lengths (θ is the scattering angle in the lab system)

---

[2]The quanta are a hopeless mess.

Fig. 3.1 Schematic representation of Compton scattering in the lab system. A photon with wave length λ is scattered on an electron at rest with scattering angle θ. The relation (3.13) determines the wave length λ' of the outgoing photon depending on λ and θ

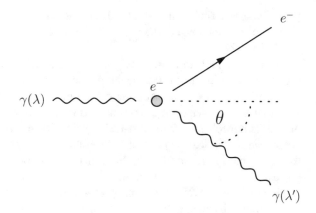

$$\lambda' - \lambda = \frac{h}{m_e c} (1 - \cos\theta) \tag{3.13}$$

cannot be explained classically (the classical result is obtained in the limit $h \to 0$), but it follows directly from energy-momentum conservation if the radiation is represented by photons. This confirmed the validity of energy-momentum conservation also in the atomic domain, which had sometimes been questioned.

Matrix and Wave Mechanics

By the early 1920s, wave-particle duality for electromagnetic radiation was established definitively. In a remarkable thesis, Louis de Broglie proposed the hypothesis that this duality should also hold for matter particles so that for instance an electron would also exhibit wave properties (de Broglie 1925). Specifically, a matter particle with momentum $\vec{p}$ would have a corresponding wave length $\lambda = h/|\vec{p}|$. Before looking closer at the motivation for this hypothesis, let us consider the magnitude of the wave length of an electron with the typical energy of 1 eV. For a nonrelativistic electron energy and momentum are related as $E = \vec{p}^2/2\,m_e$. The de Broglie wave length of an electron with energy 1 eV is therefore $\lambda \simeq 1.2\,\text{nm} = 12\,\text{Å}$. Consequently, de Broglie proposed to look for diffraction patterns in the scattering of electrons on crystals, in analogy to the diffraction of X-rays by von Laue. To be able to verify diffraction, the de Broglie wave length must be comparable to or smaller than the lattice spacings in the crystal. In 1927 Clinton Davisson and Lester Germer performed an experiment scattering electrons with an average energy of 54 eV on a nickel crystal (Davisson and Germer 1927). They could indeed detect an interference pattern where the maximal intensities occurred for scattering angles fulfilling the Bragg equation that William Bragg had originally established for the diffraction of X-rays on crystals. Thus matter waves were confirmed experimentally. In the same year, George P. Thomson (son of J. J. Thomson) and Alexander Reid also observed diffraction phenomena in the scattering of electrons on thin celluloid films (Thomson and Reid 1927).

Let us now look in more detail at the considerations of de Broglie that played a decisive role in the further development of quantum mechanics, especially for the formulation of the Schrödinger equation. It must be emphasised right from the start that the following arguments should in no way be considered as a derivation of quantum mechanics from classical physics. Quantum theory constitutes a radical breach with classical physics and the founding fathers were using analogies like the following in groping for the final theory. This procedure was originally called correspondence principle by Bohr.

However, for the following analogies – one can also call them speculations – some knowledge of basic mathematics is needed, though not going beyond the upper secondary level syllabus (see also Appendix A). According to Galilei, the book of nature is written in the language of mathematics. Einstein provided the appropriate verdict: "One should treat matters as simply as possible but not simpler." Nevertheless, the remainder of this chapter as well as the whole book should also be understandable without the following mathematical fragments.

Starting point is a classical electromagnetic wave in the vacuum, a solution of the free Maxwell equations (Appendix B). Each such wave, denoted in general as a wave packet, can be decomposed into a superposition (Fourier decomposition) of plane waves. For this reason, the plane wave is especially popular with theoreticians, described mathematically in terms of trigonometric functions (sine and cosine). Cosine and sine are also real and imaginary parts of an exponential function (Appendix A). Therefore, a plane wave can be written – up to a factor denoting the amplitude of the wave – as real or imaginary part of the function

$$\varphi(t, x) = e^{-i(\omega t - k x)} . \tag{3.14}$$

Here $\omega = 2\pi \nu$ is the angular frequency and we restrict ourselves for simplicity to one space dimension represented by the coordinate x. The wave number k (a wave vector in three dimensions) is essentially an inverse wave length: $k = 2\pi/\lambda$. In classical electrodynamics, waves in vacuum are described by real-valued functions. On the other hand, wave functions in quantum mechanics are in general complex. Therefore, it makes sense to use the complex-valued exponential function in Eq. (3.14) in the following.

We now interpret (3.14) as wave function of a photon. According to Planck and Einstein, we then have $E = h\nu = \hbar\omega = hc/\lambda$. As the photon is massless, the photon momentum is related to the energy by $p = E/c$ (special theory of relativity) and therefore

$$p = h/\lambda = h k/2\pi = \hbar k . \tag{3.15}$$

The wave function (3.14) can then also be written as

$$\varphi(t, x) = e^{-\frac{i}{\hbar}(E t - p x)} . \tag{3.16}$$

At this point de Broglie postulates in his thesis that (3.16) also holds for matter particles, with the energy to be modified appropriately (see below).

Attempting to understand atomic spectra, Wolfgang Pauli concluded that an electron with given momentum must exist in two versions. In addition to mass and charge there must therefore exist an additional degree of freedom. This leads him in 1925 to the exclusion principle (Pauli 1925): two electrons with the same quantum numbers (including the additional degree of freedom) cannot be in the same atomic state. This ad hoc principle explains the structure of the electronic shells and at the same time the stability of atoms. There cannot be more than two electrons in the ground state.

Still in the same year Samuel Goudsmit and George Uhlenbeck proposed (Goudsmit and Uhlenbeck 1925) that the extra degree of freedom postulated by Pauli is an intrinsic angular momentum (spin). However, the classical picture of the electron as a rotating sphere is untenable. Therefore, not only the unsparingly critical Pauli but also Goudsmit and Uhlenbeck were not too sure about their case. But they were encouraged by their director Paul Ehrenfest at the University of Leiden to publish their proposal (cited in Uhlenbeck 1976): "You are both young enough to be able to afford a stupidity."

In July 1925 Werner Heisenberg hands over a manuscript to his director Born. The latter characterises the paper in a letter to Einstein as follows: "Heisenbergs neue Arbeit, die bald erscheint, sieht sehr mystisch aus, ist aber sicher richtig und tief.[3]" Heisenberg starts from the assumption that unmeasurable quantities like trajectories and orbiting times of electrons in Bohr's atomic model should have no place in a new theory. Instead he considers with help of the correspondence principle relations between measurable quantities such as the frequencies of spectral lines. If one would use Heisenberg's fundamental paper (Heisenberg 1925) in an introductory course on quantum mechanics, one would probably find oneself before an empty auditorium for the following lecture. In this first paper on his so-called matrix mechanics the word matrix does not even appear. But Born immediately recognizes the underlying mathematical structure. Together with his student Pascual Jordan, a few weeks later they not only formulate matrix mechanics but also the fundamental commutation relations between position and momentum operators. The definitive formulation of matrix mechanics was then presented at the end of the year by Born et al. (1925).

How can one picture a matrix or the more general concept of an operator in quantum mechanics? The matrix associated with a specific physical quantity contains (slightly simplified) the complete set of all possible measurements of that quantity. The actual measurement then determines one of the possible values, in the language of quantum mechanics by the expectation value of the corresponding operator in a given state. In the so-called position space an operator is a prescription what to do with a function like in Eq. (3.16). For instance, the position operator X is simply represented by multiplication of the function with the coordinate x – we still work with only one spatial dimension:

---

[3] Heisenberg's new paper that will appear soon looks very mystical but is certainly correct and profound.

$$X \,\varphi(t, x) = x \,\varphi(t, x) \,. \tag{3.17}$$

The momentum operator, on the other hand, is represented in position space by the differential operator[4] $P = \dfrac{\hbar}{i} \dfrac{\partial}{\partial x}$, because (see Appendix A)

$$P \,\varphi(t, x) = \frac{\hbar}{i} \frac{\partial}{\partial x} e^{-\frac{i}{\hbar}(E t - p x)} = p \,\varphi(t, x) \,. \tag{3.18}$$

Let us now apply both X and P to an arbitrary wave function $\psi(t, x)$. Depending on the order of operations, keeping in mind the product rule for differentiation in the second formula, one obtains

$$X P \,\psi(t, x) = X \left\{ \frac{\hbar}{i} \frac{\partial}{\partial x} \psi(t, x) \right\} = \frac{x \hbar}{i} \frac{\partial}{\partial x} \psi(t, x) \tag{3.19}$$

$$P X \,\psi(t, x) = \frac{\hbar}{i} \frac{\partial}{\partial x} \{x \,\psi(t, x)\} = \frac{\hbar}{i} \psi(t, x) + \frac{x \hbar}{i} \frac{\partial}{\partial x} \psi(t, x) \tag{3.20}$$

and thus[5]

$$(X P \, - P X)\psi(t, x) = [X, P]\psi(t, x) = i \hbar \psi(t, x) \,. \tag{3.21}$$

Since the result is completely independent of the arbitrary wave function $\psi(t, x)$, we can also write (3.21) with help of the unit operator $\mathbb{1}$ as an operator relation

$$[X, P] = i \hbar \mathbb{1} \,. \tag{3.22}$$

Congratulations! You have just "derived" the fundamental commutation relations between position and momentum operators in quantum mechanics. A little bit later we will return to the consequences of this relation.

In the beginning of 1926, Pauli published his calculation of the spectrum of the hydrogen atom on the basis of matrix mechanics (Pauli 1926). Again this calculation is not necessarily suitable for an introductory course, but it was of great importance for the general acceptance of matrix mechanics. Immediately afterwards the next milestone appeared. On January 27 Erwin Schrödinger submitted his first paper on wave mechanics with the title "Quantisierung als Eigenwertproblem" (Schrödinger 1926a). In that paper not only the Schrödinger equation is introduced but it is also used to calculate the hydrogen spectrum. In comparison with the still unfamiliar matrix mechanics, Schrödinger orients himself more towards classical physics. Therefore, the approach to quantum mechanics via the Schrödinger equation is still simpler even today. After three more articles on wave mechanics, Schrödinger proves the equivalence of matrix and wave mechanics in the same year (Schrödinger 1926b). It seems that not only he had been surprised initially: "Bei der

---

[4]In the so-called momentum space the roles of X and P would have to be interchanged.
[5]The square brackets denote the antisymmetric product (commutator) of two operators.

außerordentlichen Verschiedenheit der Ausgangspunkte und Vorstellungskreise der Heisenbergschen Quantenmechanik einerseits und der neulich hier in ihren Grundzügen dargelegten und als "undulatorische" oder "physikalische" Mechanik bezeichneten Theorie andererseits, ist es recht seltsam, daß diese beiden Quantentheorien hinsichtlich der bisher bekannt gewordenen speziellen Ergebnisse miteinander auch dort übereinstimmen, wo sie von der alten Quantentheorie abweichen ...Das ist wirklich sehr merkwürdig, denn Ausgangspunkt, Vorstellungen, Methode, der ganze mathematische Apparat scheinen in der Tat grundverschieden.[6]" In the same year, Paul Dirac publishes an abstract formulation of quantum mechanics as "transformation theory" (Dirac 1927), which contains both matrix mechanics and wave mechanics as special cases. That paper also contains a proof that quantum mechanics turns into classical mechanics in the limit $\hbar \to 0$ as expected.

We can also "derive" the Schrödinger equation using the previously introduced analogies. We start from the wave function (3.16) and use the relation $E = p^2/2m$ for the energy of a free, nonrelativistic particle with mass m. As shown in Eq. (3.18), the momentum operator in position space is given by the differential operator $P = \dfrac{\hbar}{i}\dfrac{\partial}{\partial x}$ and the square of the momentum therefore by the second partial derivative $P^2 = -\hbar^2 \dfrac{\partial^2}{\partial x^2}$. Thus we can write

$$E\,\varphi(t,x) = i\,\hbar\frac{\partial}{\partial t}\varphi(t,x) = \frac{P^2}{2m}\varphi(t,x) = -\frac{\hbar^2}{2m}\frac{\partial^2}{\partial x^2}\varphi(t,x) \qquad (3.23)$$

and we have arrived at the Schrödinger equation for a free massive particle in one space dimension. Returning now to three dimensions and adding to the kinetic energy a potential $V(x, y, z)$ for the interaction, we obtain the Schrödinger equation in its general form and in full beauty:

$$i\,\hbar\frac{\partial}{\partial t}\psi(t,x,y,z) = \left\{-\frac{\hbar^2}{2m}\left(\frac{\partial^2}{\partial x^2} + \frac{\partial^2}{\partial y^2} + \frac{\partial^2}{\partial z^2}\right) + V(x,y,z)\right\}\psi(t,x,y,z)\,.$$
$$(3.24)$$

With the Laplace operator $\Delta = \dfrac{\partial^2}{\partial x^2} + \dfrac{\partial^2}{\partial y^2} + \dfrac{\partial^2}{\partial z^2}$ and with the Hamilton operator (in position space) $H = -\dfrac{\hbar^2}{2m}\Delta + V(\vec{r})$ we can write the general Schrödinger equation[7] in the usual compact form

---

[6]Given the extraordinary differences of starting points and concepts of Heisenberg's quantum mechanics on the one hand and the "wave-like" or "physical" mechanics, recently presented here with its main features, on the other hand, it is quite strange that these two quantum theories agree with each other in all presently known results even where they deviate from the old quantum theory ...This is indeed very remarkable because starting point, viewpoints, method, the whole mathematical apparatus actually seem to be completely different.

[7]In full generality the potential can also depend on time: $V(t, \vec{r})$.

$$ih\frac{\partial}{\partial t}\psi(t,\vec{r}) = H\,\psi(t,\vec{r}) \qquad (3.25)$$

where we have collected the space coordinates (x, y, z) in a position vector $\vec{r}$. This lean equation, after suitable generalisation for an arbitrary number of particles, accounts for the whole nonrelativistic quantum physics, chemical binding, large parts of solid state physics, etc. It couldn't be simpler!

The triumphal march of the Schrödinger equation was irresistible, but at the same time Pandora's box was opened. How should one interpret the mysterious wave function $\psi(t, \vec{r})$, the solution of the Schrödinger equation? Einstein, de Broglie and Schrödinger always viewed it classically as a sort of accompanying wave of the particle. While Schrödinger was convinced all his life that the "quantum theoretical oddities" were repaired by his wave equation, the statistical interpretation of the wave function prevailed more and more, an interpretation first formulated by Born (1926). The quantum mechanical wave function does not describe a physical wave, but a "probability amplitude" whose absolute square indicates the probability for finding the particle in a measurement for instance at a certain position. In Born's paper one can find a footnote that has acquired a certain fame at least among quantum engineers. On the basis of a scattering process, Born originally concluded that the wave function itself is a measure of the probability. A footnote in the publication modifies this conclusion: "Anmerkung bei der Korrektur: Genauere Überlegung zeigt, daß die Wahrscheinlichkeit dem Quadrat der Wellenfunktion proportional ist.[8]" Narrow-minded colleagues might be tempted to scoff that even a footnote may be worth a Nobel Prize. The statistical interpretation was supported and further developed especially by Bohr and his school. In his correspondence with Einstein, Born later recognized the importance of Bohr's contributions for the statistical interpretation, but he also added: "Daß sie überall als Kopenhagener Auffassung zitiert wird, scheint mir jedoch nicht gerechtfertigt.[9]" At least Born received in 1954, late but not too late, the Nobel Prize for the statistical interpretation of the wave function.

At the end of 1926 Pauli writes in a letter to Heisenberg (Pauli 1979): "Man kann die Welt mit dem p-Auge und man kann sie mit dem x-Auge ansehen, aber wenn man beide Augen zugleich aufmachen will, wird man irre.[10]" As usual with Pauli, the letter did not only contain this sentence but a whole treatise. This explains Heisenberg's reaction: " ..., daß Ihr Brief dauernd hier die Runde macht und Bohr, Dirac und Hund uns dauernd darum raufen.[11]" It seems that Heisenberg won the fight in the end because already in March 1927 he submits a paper with the title " Über den anschaulichen Inhalt der quantentheoretischen Kinematik und Mechanik" (Heisenberg 1927). In the abstract he says: " ...es wird gezeigt, daß kanonisch konjugierte

---

[8]Note added in proof: a more detailed consideration shows that the probability is proportional to the square of the wave function.

[9]However, it does not seem to be justified to me that it is generally referred to as Copenhagen interpretation.

[10]One can look at the world with the p-eye and one can look at it with the x-eye, but if you want to open both eyes at the same time you go crazy.

[11] ...Your letter keeps circulating here and Bohr, Dirac, Hund and I keep fighting over it.

Größen simultan nur mit einer charakteristischen Ungenauigkeit bestimmt werden können.[12]" Canonically conjugate quantities, in particular position and momentum, are characterised by a commutator of the form (3.22). The physical content of the famous uncertainty relation is analysed in great detail in Heisenberg's paper. On the other hand, the uncertainty relation as an inequality

$$\Delta x \, \Delta p \geq \hbar/2 \qquad (3.26)$$

as known today is not contained in Heisenberg's paper, mainly because Heisenberg does not define a precise measure for the uncertainties Δx and Δp. In common parlance, the content of the uncertainty relation is often described as "nothing is known precisely", which is about as correct as "everything is relative".

How then should one interpret the uncertainty relation? Let us assume that an experimenter prepares an experiment that allows him to measure the position and the momentum of a particle as often as he wishes, i.e., he can reproduce after each measurement (of position or of momentum) the original starting point of his experiment. For instance, he starts with a series of position measurements. Because of the statistical character of quantum mechanics the measured values will not all be the same but they will scatter around a mean value. Mathematicians have defined as measure of this scatter a quantity that in the case of a position measurement is called the mean square deviation $(\Delta x)^2$. After the position measurements our experimenter turns to the momentum measurements and will find a mean square deviation $(\Delta p)^2$. The uncertainty relation now states that, regardless how precisely the experimentalist has prepared and performed his measurements and independently how often he has repeated them, the product $\Delta x \, \Delta p$ will always be bigger than or at best equal to $\hbar/2$. For instance, he can decide to arrange his experiment in such a way as to measure very precisely the position of the particle. If he then also measures the momentum with the same setup, the mean square deviation $(\Delta p)^2$ will be correspondingly bigger such that the inequality (3.26) will always be satisfied. Giants of physics like Einstein have repeatedly tried to find counter-examples to the uncertainty relation, which in the end were always refuted mainly by Bohr.

The uncertainty relation does not only hold for position and momentum but for all operators fulfilling the commutation relation (3.22). The corresponding physical quantities (observables) are referred to as canonically conjugate. The converse is not true. One can also define a kind of uncertainty relation for energy and time[13] and yet energy and time are not canonically conjugate in the sense of (3.22). This will also play a role in Chap. 4 when trying to bring the canonical commutation relations into accordance with special relativity.

---

[12] …it is shown that canonically conjugate quantities can simultaneously be determined with a characteristic uncertainty only.

[13] See, e.g., https://iopscience.iop.org/article/10.1088/1742-6596/99/1/012002/pdf.

Schrödinger–Pauli and Dirac Equations

At the time when quantum mechanics was formulated, the measurements of spectral lines were already so precise that one observed definite deviations from the theoretical predictions for the energy levels. Schrödinger rightly suspected that this fine structure has to do with relativistic corrections to the predictions of the nonrelativistic Schrödinger equation. We have just seen when "deriving" the Schrödinger equation that the equation is nonrelativistic. One obvious feature is that the Schrödinger equation (3.24) is a differential equation of first order in the time but of second order in the spatial coordinates. Since the Lorentz transformation (B.11) mixes spatial and temporal coordinates the Schrödinger equation cannot be Lorentz invariant. There are two options for a relativistic generalisation of the Schrödinger equation: it contains either only first-order derivatives or only second-order ones. Already in 1926 Schrödinger took up the second option and found out that he could not explain the fine structure of the hydrogen spectrum in this way. He therefore put the project aside and for that reason the corresponding equation is known as Klein–Gordon equation today. The main reason for the failure was that there is no place for the electron spin in the Klein–Gordon equation. The equation is however used successfully in investigations of pionic atoms where the electron is replaced by a charged pion (with spin 0). Moreover, the Klein–Gordon equation plays a prominent role in quantum field theory, in particular for the description of the Higgs field and its quantum, the Higgs boson.

By including the spin of the electron, Pauli extended the Schrödinger equation to the Schrödinger–Pauli equation (Pauli 1927). Whereas the orbital angular momentum can only adopt integer multiples of $\hbar$ in quantum theory, the spin can also have half-integer values, in particular $\hbar/2$ in the case of an electron. The wave equation has then two components for the two spin orientations (e.g., spin up and spin down) and therefore the Schrödinger–Pauli equation is a two-component matrix equation. However, like the Schrödinger equation the Schrödinger–Pauli equation is not Lorentz invariant either. But it provides the correct explanation of the famous Stern–Gerlach experiment[14] (Gerlach and Stern 1922). In that experiment a beam of silver atoms was split up in two separate beams by traversing an inhomogeneous magnetic field. The two beams were then detected on a screen (Fig. 3.2). According to the Schrödinger–Pauli equation the relevant quantity is the magnetic moment of the atom. The silver atom has a single electron in its outermost shell (valence electron) and this electron has orbital angular momentum zero. In the physics jargon it is in an S state. Therefore, the total angular momentum of the silver atom consisting in general of both orbital angular momentum and spin is equal to the spin of the valence electron in this special case. Thus the magnetic moment is proportional to

---

[14]The experiment was performed by Stern and Gerlach in 1922, three years before the actual introduction of spin. At that time it was actually interpreted as strong support for the Bohr–Sommerfeld model. The history of the experiment and of its reception can be found in Pakvasa (2018).

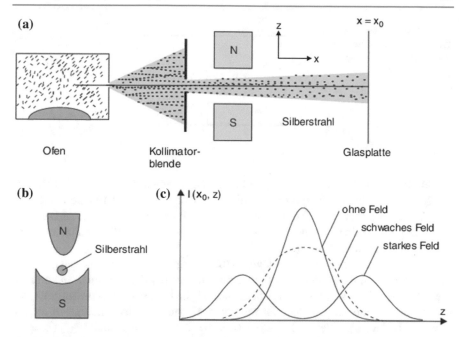

Fig. 3.2 Stern–Gerlach experiment. **a** Experimental setup; **b** Section through the inhomogeneous magnetic field; **c** Blackening intensities on the detector screen with (two peaks) and without (single peak) magnetic field (From Demtröder 2010; with kind permission of © Springer-Verlag Berlin/Heidelberg 2010. All Rights Reserved)

the electron spin. The two possible orientations of the magnetic moment for spin $\hbar/2$ then explain the splitting into two beams.

Let us return to the relativistic generalisation of the Schrödinger equation. Based on Pauli's work, Dirac constructed a relativistic wave equation (Dirac 1928) containing only first derivatives in both spatial and temporal coordinates. In doing so he noted that for a charged and therefore massive particle[15] at least four components of the wave equation were needed. The Dirac equation is therefore a four-dimensional matrix equation (Eq. (5.2)) and Dirac drew immediately the right conclusion. In addition to the two components for the electron, two more components are foreseen for the antiparticle with an opposite (positive) charge to the electron. At that time, the only candidate was the positively charged proton, which Dirac therefore initially suggested to be the antiparticle of the electron. The objection of Robert Oppenheimer followed promptly (Oppenheimer 1930). In this case the hydrogen atom would annihilate immediately and there would not be any stable atoms. In fact, soon afterwards Carl Anderson detected the correct antiparticle of the electron in cosmic rays, the positron with the same mass as the electron, but with opposite charge (Anderson 1932).

---

[15] All charged particles have nonvanishing mass.

Fig. 3.3 Fine structure of the
hydrogen atom for the
energy levels with principal
quantum number $n = 2$
(Fig. 2.1)

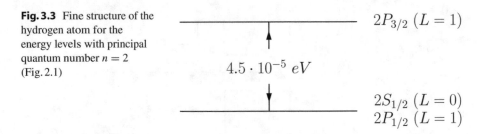

$2P_{3/2}\ (L = 1)$

$4.5 \cdot 10^{-5}\ eV$

$2S_{1/2}\ (L = 0)$
$2P_{1/2}\ (L = 1)$

Taking the electron spin into account in the Dirac equation also explains the fine structure of the hydrogen spectrum. This fine structure is a relativistic correction of order α^4 and higher (hence the name fine-structure constant for α), whereas the Schrödinger equation just like Bohr's atomic model in Eq. (3.10) predicts only the leading terms quadratic in α for the energies.

In Fig. 3.3 the fine structure is shown for the energy levels with principal quantum number $n = 2$. In the Schrödinger theory the energy only depends on n. In contrast, the Dirac equation also predicts a dependence on the total angular momentum J, the sum of orbital and intrinsic angular momenta. In the usual terminology of atomic physics, the levels are denoted as $2S_{1/2}\ (L = 0)$, $2P_{1/2}\ (L = 1)$ and $2P_{3/2}\ (L = 1)$, where the indices $1/2$, $3/2$ specify the values of J (always in units of $\hbar$). Moreover, the Dirac equation was also able to completely explain the Zeeman effect (splitting of spectral lines in a homogeneous magnetic field).

Planck's radiation formula and Einstein's photon hypothesis stood at the beginning of the quantum era. On the other hand, photons do not appear at all in modern quantum mechanics whose historical development we have sketched in this chapter. Both in the Schrödinger and in the Dirac equation only a classical electromagnetic field has its place. The main problem for a quantum theory of photons is that photons always travel with the speed of light. Therefore, any nonrelativistic approximation is doomed to failure from the very start. The only known successful method is to quantise the electromagnetic field and this approach was pursued soon after the construction of quantum mechanics. First successes of this approach, in particular the quantisation of the free electromagnetic field, were achieved in the second half of the 1920s by Dirac, Pauli, Born, Heisenberg, Jordan and others. We will return to this issue in much more detail in the following two chapters.

References

Anderson CD (1932) Science 76:238
Bohr N (1913) Phil Mag 26, 1; ibid. 476; ibid. 857
Born M (1926) Z Phys 37:863
Born M, Heisenberg W, Jordan P (1925) Z Phys 35:557
Compton AH (1923) Phys Rev 21:483

Davisson C, Germer LH (1927) Nature 119:558

de Broglie L (1925) Ann de Physique 3:22

Demtröder W (2010) Experimentalphysik 3: Atome, Moleküle und Festkörper. Springer Spektrum, Berlin

Dirac PAM (1927) Proc Royal Soc A 113:621

Dirac PAM (1928) Proc Royal Soc A 117:610

Einstein A, Born H, Born M (1969) Briefwechsel 1916–1955. Nymphenburger Verlagshandlung, München

Franck J, Hertz G (1914) Verhandlungen der Deutschen Phys Ges 16:457

Gerlach W, Stern O (1922) Z Phys 9:349

Heisenberg W (1925) Z Phys 33:879

Heisenberg W (1927) Z Phys 43:172

Oppenheimer JR (1930) Phys Rev 35:562

Pakvasa S (2018) The Stern-Gerlach experiment and the electron spin. arXiv:1805.09412

Pauli W (1925) Z Phys 31:765

Pauli W (1926) Z Phys 36:336

Pauli W (1927) Z Phys 43:601

Pauli W (1979) Wissenschaftlicher Briefwechsel mit Bohr, Einstein, Heisenberg u. a., vol I. In: Hermann A, von Meyenn K, Weisskopf V (eds) Springer, Berlin

Schrödinger E (1926a) Ann Phys 79:361

Schrödinger E (1926b) Ann Phys 79:734

Thomson GP, Reid A (1927) Nature 119:890

Uhlenbeck GE (1976) 50 years of spin: personal reminiscences. Phys Today 29:43

Uhlenbeck GE, Goudsmit SA (1925) Naturwissenschaften 13:953

Why Quantum Field Theory?

4

Causality and Quantum Fields

Since the beginning of quantum mechanics people have sometimes expressed concern that causality could be violated in quantum mechanics. There exist countless
treatises on the concept and the consequences of causality. For the simple physicist
causality is violated if information can be transmitted with velocities exceeding the
speed of light. Let us consider a free, nonrelativistic particle with mass m, momentum
$\vec{p}$ and consequently with energy $E = \vec{p}^2/2m$. It is a nice exercise for the student of
quantum mechanics to calculate the probability amplitude for the particle to get in a
given time t from a starting point $\vec{r}_0$ to an arbitrary point $\vec{r}$. The answer is

$$A(\vec{r}_0, \vec{r}; t) = \left(\frac{m}{2\pi i \hbar t}\right)^{3/2} e^{\frac{i m(\vec{r}-\vec{r}_0)^2}{2\hbar t}}. \tag{4.1}$$

Thus, there is a nonvanishing probability, namely the absolute square of the amplitude (4.1), that the particle reaches any point in space in the given time t. Thereby
information could be transferred with a superluminal velocity, albeit with a certain
probability only, and causality would be violated. One could argue that in order to
solve this problem one should better use the relativistic energy-momentum relation
$E = \sqrt{\vec{p}^2c^2 + m^2c^4}$. The result (Peskin and Schroeder 1995) differs from (4.1), but
the conclusion is the same: causality is violated.

However, surprising is not the violation of causality in quantum mechanics but
that some physicists are still surprised about this conflict with causality. Quantum
mechanics is a nonrelativistic theory (Chap. 3) even if one imposes the relativistic
energy-momentum relation. There is no reason why superluminal velocities cannot
occur in quantum mechanics. The situation reminds me of a thought experiment that
Walter Thirring used in his lectures to discuss the generation of superluminal velocities in the classical mechanics of rigid bodies. Take a very long (!) pair of scissors

© Springer Nature Switzerland AG 2019
G. Ecker, *Particles, Fields, Quanta*, Undergraduate Lecture Notes in Physics,
https://doi.org/10.1007/978-3-030-14479-1_4

Fig. 4.1 Tracks of more than 100 charged particles in the CMS detector at the LHC: proton-proton collisions at 7 TeV center-of-mass energy (With kind permission of © CERN 2010 for the benefit of the CMS Collaboration. All Rights Reserved)

and open it. If the scissors are long enough the tips will move apart with a velocity exceeding the speed of light. As bizarre as this example may seem, it occupied physicists in a more sophisticated version for some time after 1905 (Ehrenfest paradox, Ehrenfest 1909). The solution of the paradox consisted in the simple realisation that rigid bodies cannot exist in special relativity. Similarly, one has to accept that causality is violated in nonrelativistic quantum mechanics.

Before we deal with the consequences of this insight, we consider an additional argument for the necessity of extending quantum mechanics. Somewhat simplified, theoreticians say that quantum mechanics is a one-particle theory. More precisely, the framework of quantum mechanics does not allow for processes where particles are created or annihilated.

In other words, in quantum mechanics one can very well study the elastic scattering of two particles. But inelastic processes that occur regularly in accelerators like the Large Hadron Collider LHC, where in the collisions of two protons hundreds of particles are created (Fig. 4.1), are not accessible in quantum mechanics as a matter of principle. The crucial quantity is the available energy in the collision. As soon as that energy satisfies the condition $E > 2\,mc^2$, a particle with mass m and its antiparticle can be produced.[1] Such scattering processes, let alone particle decays, require a relativistic multi-particle theory in contrast to the nonrelativistic one-particle theory quantum mechanics.

---

[1]Particles that are identical with their antiparticles such as the photon or the neutral pion π^0 can also be produced singly.

The canonical commutation relations (3.22) (generalised to three space dimensions) are also incompatible with special relativity. Changing the inertial system, time and space coordinates are transformed into each other as are energy and momentum (Appendix B). In a first step, one could therefore try to introduce in addition to position and momentum as in (3.22) also operators for time and energy. We have encountered such an operator for the energy in the form of the Hamilton operator H in the Schrödinger equation (3.25). An operator for time has not shown up yet, but maybe it is worth a try, let us call it T. Compatibility with special relativity requires then (at least) that also H and T obey the canonical commutation relation

$$[H, T] = i \hbar \mathbb{1} . \tag{4.2}$$

But now we have arrived at an impasse. Position and momentum, whose operators satisfy the commutation relation (3.22), can adopt any real value between $-\infty$ and $+\infty$ and there are no discrete position or momentum eigenvalues. It can be shown that these properties are direct consequences of the canonical commutation relation (3.22). Therefore, energy and time would have to have the same properties. We would not have a problem with time in this respect but for the energy this would be a catastrophe. The energy would not be bounded from below, i.e., there would not be a lowest energy eigenvalue (instability!), and there would not exist any discrete energy levels in contradiction to significant achievements of quantum mechanics.

So we really seem to be stuck in a dead end. The relativistic invariance of the theory requires the same status both for position and time and for momentum and energy but the four-dimensional form of the commutation relations leads to blatant contradictions. The only way out seems to be to deprive the spatial coordinates of their operator status and to downgrade them together with time to normal nonquantised space-time variables. The quantum object that we need of course in a quantum theory will then depend on those space-time coordinates. The simplest possibility is that the quantum objects depend on a single space-time point only. Such objects are called (local) quantum fields. In a relativistic notation, time and space coordinates are combined in a four-vector $x = (ct, \vec{r})$. The fields (field operators) of relevance for particle physics are collected in Table 4.1.

In the third column of Table 4.1 only a few representative examples are listed but the particle zoo of course is much bigger (Chap. 9). However, the list of field types is complete according to the present state of knowledge. In other words, the Standard

Table 4.1 Relevant types of fields in the Standard Model of particle physics. From now on we denote the spin of a particle simply as spin s instead of the explicit $s \hbar$, i.e., the factor $\hbar$ is always implied

Field	Type	Representative particle	Spin
$\varphi(x)$	Scalar field	Higgs boson	0
$\psi(x)$	Spinor field	Electron	1/2
$A(x)$	Vector field	Photon	1

Model of particle physics contains only scalar, spinor and vector fields. A tensor field (spin 2) with the hypothetical graviton as particle would require a quantum field theory of gravity which we still do not have. Fields with spin-3/2 particles occur in some extensions of the Standard Model such as in supergravity.

A few supplementary remarks are listed here.

   i. Incorporating both particle and wave aspects in a quantum field is therefore not specific for the electromagnetic field, but it applies for all fundamental particles (quanta): matter particles (leptons and quarks), interaction quanta (photons, W- and Z bosons, gluons), Higgs boson.
  ii. Heisenberg's commutation relations are replaced in quantum field theory by commutation or anticommutation relations for quantum fields. Instead of the less appropriate designation "second quantisation" – in distinction to the first quantisation in quantum mechanics – we speak of field quantisation. Although not strictly deducible from quantum mechanics, the step from quantum mechanics to quantum field theory is a small step in comparison with the revolutionary transformation from classical physics to quantum mechanics.
 iii. In contrast to quantum mechanics causality is guaranteed in relativistic quantum field theories. This appears plausible because all ingredients of the theory are in accordance with special relativity, but for an actual proof we have to refer to pertinent textbooks (e.g., Peskin and Schroeder 1995).
  iv. Relativistic invariance demands a departure from simple quantum mechanics but local quantum field theory is only the simplest solution. Dirac seems to have been the first in 1962 to investigate the quantisation of extended objects. In the following decade string theory was born where one-dimensional objects (strings) are the fundamental quantities. Especially as superstring theory this approach has almost become a separate branch of science that in the meantime does not even shy away from higher-dimensional objects (membranes). However, so far nature seems to be satisfied with quantum field theory as the basis of the Standard Model. In any case, up to now there is not a single experimental hint that would support more exotic avenues.

Spin and Statistics

In June 1924, i.e. even before the birth of quantum mechanics, Einstein receives a letter from India. A young physicist by the name of Satyendranath Bose asks Einstein to review the enclosed article "Planck's law and the hypothesis of light quanta" and to submit the paper for publication (Bose 1924). In this paper Bose derives the radiation law starting from the ad hoc assumption that unlike in the classical Maxwell-Boltzmann statistics photons are indistinguishable particles. Einstein immediately recognized the importance of the paper that also contains the first quantum mechanical formulation of an ideal (Bose) gas. In addition, Einstein generalised Bose's ansatz for particles with nonvanishing mass and predicted that a part of the ideal gas should condense at low temperatures in the quantum mechanical ground state (Bose-Einstein condensation, Einstein 1924, 1925). This prediction was experimentally verified only in 1995.

The essential property of particles satisfying Bose-Einstein[2] statistics can already be seen with the quantum mechanical wave function of two such particles. This wave function $\psi(\vec{r}_1, s_1; \vec{r}_2, s_2)$ is symmetric under the exchange of both spatial and spin coordinates of the two particles:

$$\psi(\vec{r}_2, s_2; \vec{r}_1, s_1) = \psi(\vec{r}_1, s_1; \vec{r}_2, s_2) \,. \tag{4.3}$$

The symmetry property (4.3) cannot be valid for all particles because it contradicts in particular Pauli's exclusion principle according to which two electrons cannot have the same quantum numbers. Independently of each other, Enrico Fermi and Paul Dirac therefore proposed an alternative statistics (Fermi 1926; Dirac 1926) known as Fermi-Dirac statistics since then.[3] Particles satisfying this statistics are called fermions, with the electron as most prominent member. In contrast to (4.3), the wave function of two fermions is antisymmetric:

$$\psi(\vec{r}_2, s_2; \vec{r}_1, s_1) = -\psi(\vec{r}_1, s_1; \vec{r}_2, s_2) \,. \tag{4.4}$$

This condition implies the exclusion principle. If the two fermions have the same position and spin coordinates, i.e. $\vec{r}_1 = \vec{r}_2 = \vec{r}$ and $s_1 = s_2 = s$, then the condition (4.4) amounts to $\psi(\vec{r}, s; \vec{r}, s) = -\psi(\vec{r}, s; \vec{r}, s)$ and therefore $\psi(\vec{r}, s; \vec{r}, s)$ must be identically zero. Two fermions with the same quantum numbers cannot be in the same state. This has far-reaching consequences that go beyond atomic structure. For instance, unlike the Bose-Einstein condensate there cannot be a Fermi-Dirac condensate because the fermions cannot accumulate in the ground state. Another consequence is the band structure of solids that is based on the Fermi distribution of an electron gas.

At this point there are at least two questions.

i. Can the theory explain which particles are bosons and which are fermions?
ii. Are there additional possibilities or do Bose-Einstein and Fermi-Dirac statistics encompass all particles?

To answer these questions, we return to quantum field theory. First attempts for quantising the free radiation field can be found in the paper on matrix mechanics

---

[2]Following Dirac's suggestion, such particles are called bosons.

[3]At the end of 1925 Jordan presented a paper to Born asking for publication in Zeitschrift für Physik where Born was the editor. Shortly afterwards, Born went on an extended trip to the U.S. and forgot about the paper he had stowed away in his suitcase. Born later (English translation by the author): "I hate Jordan's politics but I can never make up for what I did to him. ...In the meantime the Fermi-Dirac statistics had been discovered independently by Fermi and Dirac. But Jordan was the first." (Cited in Ehlers and Schücking 2002) Incidentally, Jordan had called his discovery Pauli statistics.

by Born, Heisenberg and Jordan (Chap. 3). But the first comprehensive treatment of the quantisation of the radiation field is contained in the paper of Dirac entitled "The Quantum Theory of the Emission and Absorption of Radiation" (Dirac 1927). In that paper the name quantum electrodynamics (QED) appears for the first time. In the following year, Pauli and Jordan formulate the commutation relations for field operators that take the place of the quantum mechanical commutation relations (3.22) (Jordan and Pauli 1928). In the same year, Jordan and Eugene Wigner also establish the quantisation conditions for Fermi fields where anticommutation rules replace the bosonic commutation relations (Jordan and Wigner 1928). To conclude this first heroic phase of quantum field theory, in 1929 Heisenberg and Pauli present the general theory of relativistic quantum fields (Heisenberg and Pauli 1929). This so-called canonical quantisation of fields is a standard method until today. These papers also address the problem that through the interaction with the radiation field the electron acquires an arbitrarily large self-energy. The divergence problem of QED will occupy the quantum field theoreticians for quite a while and we will come back to it extensively in Chap. 6.

At the end of the 1920s it was clear that photons obey Bose-Einstein statistics while electrons satisfy the Fermi-Dirac statistics. Moreover, no possibility for a different statistics emerged in the framework of quantum field theory. Nevertheless, it took another ten years till a final proof of the spin-statistics theorem that could explain the experimental findings. On the basis of relativistic quantum field theory only, Markus Fierz and Pauli proved that all particles are either bosons or fermions (Fierz 1939; Pauli 1940). The spin of the particle turns out to be crucial: particles with integer spin (always in units of $\hbar$) are bosons while those with half-integer spin are fermions.[4] The theorem not only applies to fundamental particles (quanta) but also to bound states. Today we understand all hadrons, i.e. all "particles" with strong interactions (Chap. 8), as bound states of quarks and gluons. Baryons like protons and neutrons are to a first approximation bound states of three quarks. Since the quarks have spin $1/2$, baryons have necessarily half-integer spin[5] and are therefore fermions. On the other hand, mesons like the pion are quark-antiquark bound states and are therefore bosons because they have integer spin.[6] But the spin-statistics theorem also extends to atomic nuclei as bound states of protons and neutrons. Nuclei with an odd number of nucleons (collective term for protons and neutrons) are fermions, those with an even number are bosons. The spin-statistics theorem is a cornerstone of quantum field theory and thus of the Standard Model. All experimental results of the past 90 years support it. Finally, it should be mentioned that it is essential for the proof of the theorem that we obviously live in a world with three spatial dimensions. In two space dimensions that can be relevant in solid-state physics (e.g., in the quantum Hall effect) also quasi-particles can exist, which are neither bosons nor fermions and which are called anyons (Wilczek 1991).

---

[4]Group theory tells us that there are no other possible values for angular momentum but integer or half-integer multiples of $\hbar$.

[5]Neither the gluons nor a possible orbital angular momentum can make a difference.

[6]Particles and antiparticles have the same spin.

Symmetries and Conservation Laws

Regular structures have always been fascinating. Already the ancient Greeks knew bodies with maximal symmetry. The five Platonic solids (tetrahedron, cube, octahedron, dodecahedron and icosahedron) are completely bordered by congruent regular polygons. Kepler discovered the laws of planetary orbits in trying to complete the harmony of the celestial spheres. In his book *"Mysterium Cosmographicum"* published in 1596, Kepler attempted to relate the five planets known until then (without earth), Mercury, Venus, Mars, Jupiter and Saturn, to the surfaces of the five Platonic solids.

In the 19th century the modern view of symmetries in physics evolves as the group of space-time transformations that leave Newton's equations of motion unchanged. In the last sentence the word "group" is not accidental. Symmetry transformations form a group in the mathematical sense where the essential property of a symmetry group is that the result of two successive transformations is again a symmetry transformation. A simple example are the spatial translations. Two successive translations may of course be replaced by a single one.

What may appear at first sight as child's play with Platonic solids has actually profound consequences for physics. According to a theorem of the mathematician Emmy Noether, each symmetry leads to a conservation law (Noether 1918). In classical physics this is only true for continuous symmetry transformations, roughly speaking those transformations that can be pictured as being made up of (arbitrarily) many (correspondingly) small transformations. The spatial translations mentioned above may serve as an example. A spatial translation can be thought of as consisting of many small translations. Newtonian mechanics possesses ten independent continuous symmetries. Together with the corresponding conserved quantities following from Noether's theorem, they are put together in Table 4.2. If one replaces Galilei transformations by Lorentz transformations (Appendix B), one obtains the symmetries of special relativity. These symmetries of Maxwell's electrodynamics and of relativistic mechanics correspond again to ten independent transformations yielding ten conservation laws. The group of these symmetry transformations is called Poincaré group or inhomogeneous Lorentz group.

With a few semantic modifications (conserved quantities correspond to symmetry operators), those relationships also hold in quantum theory. However, in quantum mechanics two new aspects arise.

Table 4.2 Symmetries and conserved quantities of classical mechanics

Symmetry	Number of parameters	Conserved quantity
Temporal translation	1	Energy
Spatial translations	3	Momentum
Rotations	3	Angular momentum
Galilei transformations	3	Center of mass

i. Also discrete symmetry transformations can lead to conserved quantities. Different from the previously mentioned continuous transformations, discrete symmetries cannot be put together with many small transformations. An important example is the space reflection or parity transformation $\vec{r} \rightarrow -\vec{r}$ (time remains unchanged). In quantum mechanics there exists a symmetry operator P (P for parity) whose eigenvalues can only be ± 1 because two successive parity transformations lead back to the original state. For instance, one can characterise atomic energy levels by assigning positive or negative parity to the corresponding states. This distinction is important for the understanding of selection rules for atomic transitions.

ii. For some symmetries the order of successive transformations matters. For instance, if one carries out two successive rotations around two different axes, the result will depend on the order of rotations. The corresponding symmetry groups like the group of rotations are called non-abelian groups. In contrast, the translations form an abelian group. It does not matter in which order one performs two translations. The existence of non-abelian symmetry groups leads in quantum theory to a phenomenon unknown in classical physics, the so-called degeneracy of energy levels. This phenomenon can again be elucidated for the rotation group. If the system under consideration, e.g., simply an atom, remains unchanged under rotations (rotational invariance), each energy level is characterised by a definite total angular momentum J consisting in general of orbital and intrinsic angular momenta. The corresponding magnetic quantum number m (not to be confounded with a mass) can acquire $2J + 1$ values $-J, -J + 1, \ldots, J - 1, J$. The energy level with angular momentum J then consists not of just one but of $2J + 1$ states, all with the same energy (degeneracy). But how can one distinguish those degenerate states experimentally if they all have the same energy? The simplest way is to apply a homogeneous magnetic field that points in a certain direction. The system is now no longer rotationally invariant because the magnetic field singles out a certain direction. And in fact the level splits up in $2J + 1$ equidistant levels and this splitting is known as Zeeman effect (Chaps. 2, 3). Understanding this phenomenon also implies that slightly broken symmetries (as in this case by a magnetic field) may often be detected more easily than exact symmetries.

In particle physics two more discrete symmetry transformations play a role. The Newtonian equations of motion remain unchanged under a transformation $t \rightarrow -t$ if the potential depends only on the spatial coordinates, i.e., if the potential is time independent. This implies that every solution of the equations of motion is again a solution if one changes the sign of the time coordinate. One speaks of time reversal or reversal of motion. In quantum field theory this symmetry transformation is implemented by a time reversal operator[7] T. However, unlike parity this operator does not have definite eigenvalues, but T implies relations between processes where

---

[7]Not to be mixed up with the hypothetical operator for the time itself that we investigated at the beginning of this chapter.

Table 4.3 Impact of parity P and time reversal T on some measurable quantities (observables)

Observable		P	T
Momentum	$\vec{p}$	$-\vec{p}$	$-\vec{p}$
Angular momentum	$\vec{J}$	$\vec{J}$	$-\vec{J}$
Electric field	$\vec{E}$	$-\vec{E}$	$\vec{E}$
Electric dipole moment	$\vec{J} \cdot \vec{E}$	$-\vec{J} \cdot \vec{E}$	$-\vec{J} \cdot \vec{E}$

initial and final states are interchanged. For instance, time reversal relates the amplitudes for the scattering processes $A + B \rightarrow C + D$ and $C + D \rightarrow A + B$. Using the correspondence principle, the effect of P and T on some measurable quantities can be deduced from classical physics. A few of those relations are collected in Table 4.3.

The last line in Table 4.3 implies that in theories with parity and time reversal as valid symmetries particles cannot have a permanent electric dipole moment. Up to now all experimental attempts to detect such electric dipole moments, which can be performed with great precision especially for neutrons and electrons, have produced negative results although both P and T are violated by the weak interactions. Weak means really weak in this case.

The symmetry transformations considered so far are all space-time transformations. For a reason soon to become evident, we consider here yet one more discrete transformation that interchanges particles and antiparticles but leaves the space-time coordinates unchanged. This transformation is called charge conjugation C. At first, it seems to have a realistic chance to acquire the status of a symmetry because particles and antiparticles do not only have the same absolute value of the electric charge but also the same mass.

Till the middle of last century the opinion prevailed that the symmetries discussed so far reflect the underlying simplicity of nature. Thus, there was a considerable shock when in 1957 a definite violation of parity was found in the β decay of polarised cobalt nuclei. Not only that but the weak interactions that are responsible for β decays also violate charge conjugation. Two decades later Steven Weinberg posed the following question in his Nobel Lecture of 1979: "Is nature only approximately simple?" Of course, one can also share the opinion of Tsung-Dao Lee, another Nobel Prize winner, that also approximate symmetries may indicate fundamental properties of nature as in the case of the weak interactions.

We will return to the symmetry properties of the weak interactions in Chap. 7. To conclude this chapter, we consider another general theorem of relativistic quantum field theory that has a status comparable to the spin-statistics theorem. The CPT theorem of Gerhart Lüders, John Bell, Wolfgang Pauli and Bruno Zumino asserts that every local Lorentz invariant quantum field theory is also invariant under the combined transformation CPT (Lüders 1954; Bell 1955; Pauli 1955; Lüders and Zumino 1958). An important consequence of this theorem is that particle and antiparticle must have the same mass and lifetime even if charge conjugation C is not a valid symmetry as in the weak interactions. Up to this day no violation of this fundamental theorem has been observed. On the contrary, when in 1964 also a violation of CP was detected in some decays of neutral K mesons, most particle physicists were convinced that in processes of the weak interactions also time reversal T must be violated so that the combined transformation CPT could again be a symmetry of the weak interactions. This could indeed be verified in 1999 in an experiment at CERN. Violations of the CPT theorem and/or of the spin-statistics theorem would require a radical change of the theoretical basis of fundamental physics.

References

Bell JS (1955) Proc R Soc A 231:479
Bose S (1924) Z Phys 26:178
Dirac PAM (1926) Proc R Soc A 112:661
Dirac PAM (1927) Proc R Soc A 114:243
Ehlers J, Schücking E (2002) Aber Jordan war der Erste. Phys J 1:71
Ehrenfest P (1909) Phys Zeits 10:918
Einstein A (1924) Sitzungsb Pr Akad Wiss, phys-math Kl, 261
Einstein A (1925) Sitzungsb Pr Akad Wiss, phys-math Kl, 3
Fermi E (1926) Rend Lincei 3:145
Fierz M (1939) Helv Phys Acta 12:3
Heisenberg W, Pauli W (1929) Z Phys 56:1; ibid. 59:168
Jordan P, Pauli W (1928) Z Phys 47:151
Jordan P, Wigner E (1928) Z Phys 47:631
Lüders G (1954) K Dan Vidensk Selsk, Mat-fys Medd 28:1
Lüders G, Zumino B (1958) Phys Rev 110:1450
Noether E (1918) Nachr K Ges Wiss Göttingen, math-phys Kl, 235
Pauli W (1940) Phys Rev 58:716
Pauli W (1955) In: Pauli W, Rosenfeld L, Weisskopf V (eds) Niels Bohr and the development of physics. McGraw-Hill, New York
Peskin ME, Schroeder DV (1995) An introduction to quantum field theory. Addison-Wesley, Reading
Wilczek F (1991) Anyons. Sci Am 264:58

Quantum Electrodynamics: Prototype of a Quantum Field Theory

5

Foundations of Quantum Electrodynamics

As the name indicates, quantum electrodynamics (QED) is the quantised counterpart of classical electrodynamics and it describes the interaction of the electromagnetic field with charged matter particles. In the Standard Model (Chap. 9) the matter particles are leptons and quarks. Here we restrict ourselves to the simplest and historically most relevant case of the matter particles and only deal with electrons and, as always in quantum field theory, their antiparticles, the positrons.

Since QED has a classical limit, one could fall back on the correspondence principle for its construction. The classical electric and magnetic fields are replaced by quantum fields (field operators), but which quantities should adopt the role of the classical charge and current densities in Maxwell's equations (Appendix B)? To answer this question, we refer to the continuity equation in classical electrodynamics. If the electric charge in a given volume changes, this can only happen if an electric current flows into or out of the volume. The continuity equation governs this balance and it is an essential part of classical electrodynamics. As we restrict ourselves here to electrons as matter particles (and fields), charge and current densities must be constructed from the Dirac field $\psi(x)$ for the electron (Table 4.1) so that the continuity equation also holds in QED. In the relativistic formulation the notion current conservation instead of continuity equation is usually employed. It now turns out that there is only one possibility to construct a conserved current out of the electron field.

In classical physics the theory is defined by Newton's equations of motion (A.2) for mechanics and by Maxwell's equations (B.1) for electrodynamics. But especially in mechanics it often is of advantage to use the so-called Lagrange function. With the help of that function the physicist can derive the equations of motion. The use of the Lagrange function has many advantages. In practically all cases it is more compact than the explicit equations of motion and the symmetries of a problem can

© Springer Nature Switzerland AG 2019
G. Ecker, *Particles, Fields, Quanta*, Undergraduate Lecture Notes in Physics,
https://doi.org/10.1007/978-3-030-14479-1_5

be read off more easily from the Lagrange function. While in classical physics the Lagrange function simplifies the life of the theoretician, the analogue in quantum field theory is practically indispensable, e.g., for the formulation of the Standard Model. This analogue is actually an operator-valued Lagrange density but particle physicists commonly refer to it simply as the Lagrangian. If one would write down the field equations of the Standard Model of fundamental interactions explicitly, even an expert could easily lose track, especially concerning the part of the weak interactions. For each field (each particle) there is a separate field equation but there is only one Lagrangian for all those fields, which greatly simplifies matters.

The Lagrangian of QED has the relatively "simple" form[1]

$$\mathcal{L}_{\text{QED}}(x) = \overline{\psi}(x) \left(i\, \gamma^{\mu} \left[\partial_{\mu} - i\, e\, A_{\mu}(x) \right] - m \right) \psi(x) - \frac{1}{4} F^{\mu\nu}(x) F_{\mu\nu}(x) \,. \quad (5.1)$$

For the start, the gentle readers are kindly asked to accept the expression (5.1) simply as a physics piece of art. After all, all the interaction between photons and electrons is contained in this one line! More precisely, the interaction is contained only in the term proportional to the elementary charge[2] e. Let us first consider the hypothetical case that the field $\psi(x)$ represents an uncharged particle, i.e., we put $e = 0$ in (5.1). Then the Lagrangian (5.1) falls into two pieces that have nothing to do with each other. The first term describes the free Dirac field for a particle with mass m. Hardly anybody should be surprised that the corresponding field equation is just the Dirac equation[3]:

$$\left(i\, \gamma^{\mu} \partial_{\mu} - m \right) \psi(x) = 0 \,. \quad (5.2)$$

The second term with the field $F_{\mu\nu}(x)$ describes the free electromagnetic field, in particular free photons. In the relativistic formulation, $F_{\mu\nu}(x)$ comprises the more familiar electric and magnetic fields $\vec{E}(x)$, $\vec{B}(x)$. $F_{\mu\nu}(x)$ can also be expressed by derivatives of the vector field $A_{\mu}(x)$ ($F_{\mu\nu} = \partial_{\mu} A_{\nu} - \partial_{\nu} A_{\mu}$) and this brings us back to the complete QED Lagrangian (5.1). Whereas in classical electrodynamics one may in principle ignore the potential field $A_{\mu}(x)$ by writing the Maxwell equations directly for the physically relevant fields $\vec{E}(x)$ and $\vec{B}(x)$ (Appendix B), the field $A_{\mu}(x)$ has a more fundamental meaning in QED. It is the quantum field for photons and it is a vector field because the photons have spin one.

---

[1]One reason for this "simplicity" is that we use here, other than elsewhere in the book, the usual convention of particle physicists expressing actions in units of $\hbar$ and velocities in units of c. In a notation that has the tendency to confuse the layman: $\hbar = c = 1$. ∂_{μ} stands for the partial derivative $\partial / \partial x^{\mu}$.

[2]For a particle with charge q, e.g., for a quark with charge $q = 2\,e/3$, the electron charge $-e$ must be replaced by q.

[3]γ^{μ} ($\mu = 0, 1, 2, 3$) are the four-dimensional Dirac matrices, corresponding to the four components of the field $\psi(x)$. In (5.1) and (5.2) we use the summation (or Einstein) convention where a sum over twice occurring indices is understood: e.g., $\gamma^{\mu} \partial_{\mu}$ stands for $\sum_{\mu=0}^{3} \gamma^{\mu} \dfrac{\partial}{\partial x^{\mu}}$.

Without any detailed investigations, important symmetry properties of QED can be read off directly from the Lagrangian (5.1). We will discuss two special examples more explicitly below. First of all, the Lagrangian (5.1) remains unchanged if one submits both fields and space-time coordinates to a Lorentz transformation (Appendix B). Moreover, the Lagrangian (5.1) does not single out any space-time point and thus QED is translation invariant just like the classical Maxwell theory. Altogether, QED is therefore invariant under Poincaré transformations (Chap. 4). In addition, the QED Lagrangian remains invariant also under the discrete transformations P (parity, space reflection) and C (charge conjugation). Because the CPT theorem holds in any local Lorentz invariant quantum field theory (Chap. 4), QED is necessarily also invariant under time reversal. Although not relevant for QED, one should be a bit more precise here. All symmetries mentioned so far leave the Lagrangian and thus also the field equations of QED invariant. But as the quantum field theories for the nuclear forces will demonstrate, this does not automatically imply that the solutions of the field equations share this property. This phenomenon of spontaneous symmetry breaking specific to quantum field theories does not arise in QED, but we will have to return to it in the discussion of the weak interaction.

Another internal symmetry[4] of QED needs a more elaborate discussion. In quantum mechanics, the wave function of the system in question can be submitted to a phase transformation. This transformation $\psi(x) \rightarrow e^{i\,\alpha}\psi(x)$ with a real number α has no physical consequences because only the absolute square of the wave function $\psi(x)$ is physically relevant (Chap. 3). This phase transformation also leaves the QED Lagrangian (5.1) unchanged where $\psi(x)$ now again is the Dirac field. This is because $\overline{\psi}(x)$ is essentially the operator equivalent of the complex conjugate Dirac field so that the phase α drops out in (5.1). The corresponding conserved quantity (Noether theorem, Chap. 4) is the electron number (more generally, the fermion number), which implies that in each QED process (more generally, in each process of the Standard Model) the number of electrons (fermions) minus the number of positrons (antifermions) does not change.

But there is still more to the symmetries of QED. We now modify the phase transformation just discussed by letting the phase depend also on the coordinates, i.e., we consider now a transformation $\psi(x) \rightarrow e^{i\,e\,\beta(x)}\psi(x)$ with a real function $\beta(x)$. Under this transformation the Lagrangian (5.1) changes because there is a space-time derivative in the first term. This modification can be compensated by adding a term $\partial_\mu\beta(x)$ to the vector field $A_\mu(x)$. Such a change is well known in classical electrodynamics. It is a so-called gauge transformation on the potential $A_\mu(x)$ that leaves the physically relevant electromagnetic fields $\vec{E}(x)$, $\vec{B}(x)$ unchanged (Appendix B). In Maxwell's theory this gauge freedom often is used to simplify the solution of a physical problem. In QED this gauge freedom has still another significance. Originally, the photon field $A_\mu(x)$ has four components (three space-like and one time-like, analogous to the space-time coordinates). On the other hand, the

---

[4]Under an internal symmetry transformation only the fields but not the space-time coordinates are transformed.

photon like any other massless particle with spin $\neq 0$ has only two possible spin orientations (Wigner 1939; Peskin and Schroeder 1995) and therefore only two physical components. In classical electrodynamics these two components correspond to electromagnetic waves with left- or right-handed circular polarisations. The gauge freedom in QED guarantees that a photon with given momentum always has only two physical degrees of freedom. Therefore, QED is a so-called gauge theory with gauge group $U(1)$ where U stands for a unitary group and 1 for the only gauge function $\beta(x)$. In contrast to the symmetries considered so far, like for instance the Poincaré group, we also speak of a local symmetry transformation in this case because the transformation function $\beta(x)$ depends on the local coordinates. Gauge symmetry and local symmetry are synonymous notions.

By means of the correspondence principle gauge invariance virtually comes down from heaven and one may ask why we make such a fuss about it. For the quantum field theories of the strong and weak nuclear forces where no correspondence principle will be available, it will turn out that gauge invariance is an essential construction principle. The quanta of the corresponding vector fields that always come together with gauge symmetries will be the W and Z bosons as carriers of the weak interactions and the gluons as carriers of the strong interactions. Apparently, gauge invariance is a fundamental ingredient of all quantum field theories relevant in the microcosm.

S-Matrix and Perturbation Theory

In quantum mechanics the wave function as solution of the Schrödinger equation contains the complete physical information for the system under consideration. As mentioned in Chap. 4, because of the possibility of creation and annihilation of particles the situation in relativistic quantum field theories is much more complex. In practice, i.e. for comparing theoretical predictions with experimental results, the scattering matrix (S-matrix for short) of John Wheeler and Werner Heisenberg is the relevant quantity (Wheeler 1937; Heisenberg 1943). A specific element of the S-matrix contains the probability amplitude for the transition from a given initial state to a certain final state. Since the number of possible initial and final states is in principle unlimited, the S-matrix is an infinite-dimensional matrix. However, in practice only a few columns of this matrix are of interest, with either one particle (decay) or two particles (scattering process) in the initial state, e.g., one photon and one electron for Compton scattering (see below). If one has calculated a certain S-matrix element, its absolute square determines (as in quantum mechanics) experimentally accessible quantities like the partial decay rates of a particle or the cross section for a scattering process. The S-matrix is a so-called unitary matrix. The sum of the absolute squares of the matrix elements in one column (same for a row) equals one because this column contains all possible final states for the given initial state. Thus, the S-matrix takes into account the conservation of probability and also in this respect it is the analogue of the quantum mechanical wave function.

The S-matrix is constrained by the symmetries of the underlying quantum field theory. For instance, from the Poincaré invariance of the theory it follows that initial and final state must have the same total momentum, the same total energy and also the same total angular momentum. Current conservation of QED guarantees the conservation of electric charge as in classical electrodynamics. Therefore, in each decay and in each scattering process the total charge of the particles involved is the same in the initial and in the final state.

Now comes the big disappointment. Neither for QED nor for the Standard Model altogether a single nontrivial, exact S-matrix element is known. Nontrivial in this connection means that one really takes the interaction seriously. In the case of no interaction ($e = 0$ in the Lagrangian (5.1)) the exact solution is known: the S-matrix is the unit matrix and nothing happens at all. Therefore, some mathematical physicists have been suspecting that QED and the Standard Model do not "exist" in the mathematical sense. This conflict brought the development of quantum field theory almost to a standstill in the 1930s. Today we believe we know better where the problem is. A quantum field theory like QED assumes implicitly that the theory remains valid at arbitrarily small distances and therefore also for arbitrarily high energies. Even for QED this is not the case because of the unification with the weak interactions in the electroweak theory (Chap. 7), let alone the expected influence of quantum gravity at very small distances (Appendix A). Most particle physicists therefore view QED and the Standard Model altogether as so-called effective quantum field theories, valid only up to some definite energies. In Chap. 11 we will return to the current paradigm of effective quantum field theories.

Irrespective of the current paradigm, how is it possible to find a concrete prediction of the theory? The magic word is perturbation theory that allows to calculate S-matrix elements as power series in the relevant coupling constants. In QED we have a single coupling constant, the elementary charge e. Since according to present knowledge both photon and electron are stable particles, all physically relevant S-matrix elements in QED have exactly two particles in the initial state (scattering processes).

The calculation of S-matrix elements can be visualised by means of the famous Feynman diagrams. Those diagrams consist of vertices and lines where the vertices represent the local interaction and the lines the particles involved. In QED there is only a single vertex because the Lagrangian (5.1) only contains a single interaction term. This is another great advantage of the Lagrangian. The particle physicist can immediately read off the interaction vertices of the theory. The only interaction term of QED contains exactly three fields (particles): two electrons ($\overline{\psi}(x)$, $\psi(x)$) and one photon ($A_\mu(x)$). The fundamental vertex diagram of QED is therefore of the form given in Fig. 5.1. This vertex diagram is an integral part of every Feynman diagram in QED but it does not describe a physical process by itself. An electron cannot turn into an electron and a photon because the conservation of energy and momentum would require a zero photon energy. But a photon with zero energy and therefore also with vanishing momentum ($|\vec{p}| = E/c$ for a massless particle) is no photon at all and thus the picture in Fig. 5.1 does not represent a physical process. This is a first warning that in general Feynman diagrams should not be interpreted as actual

Fig. 5.1 Fundamental vertex diagram of QED; the full lines stand for an electron (or positron), the wavy line for a photon

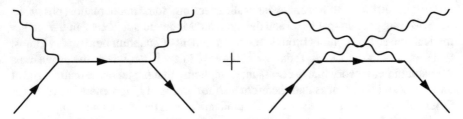

Fig. 5.2 Feynman diagrams for Compton scattering in lowest-order perturbation theory. The diagrams should be read from left (initial state) to right (final state)

processes in space and time. Feynman diagrams are suggestive prescriptions for the particle physicist how to calculate S-matrix elements for the transition from a given initial state to a well-defined final state. Feynman diagrams can be considered as a sort of construction kit where each vertex and each line stands for a definite mathematical expression. Putting the building blocks together then yields the S-matrix element in question.

In QED there are only a few initial states of practical relevance: $e^- e^-$ (two electrons), $e^- e^+$ (electron, positron), γe^- (photon, electron) and $\gamma \gamma$ (two photons). An interesting example is the elastic scattering of photons and electrons, the so-called Compton scattering $\gamma e^- \to \gamma e^-$. In lowest-order perturbation theory (also known as Born approximation), i.e. to lowest order in the unit charge e, the two Feynman diagrams in Fig. 5.2 stand for the corresponding scattering amplitude (S-matrix element).

Since each diagram contains two vertices, the corresponding S-matrix element is proportional to e^2 and thus to the fine-structure constant α. The left diagram in Fig. 5.2 suggests the interpretation that the electron in the initial state absorbs a photon, then continues on for a while before emitting finally another photon. Electron and photon in the final state are then registered in a detector. The problem with this interpretation is that the electron in the intermediate state cannot be a physical electron as we just discussed in connection with the fundamental vertex diagram in Fig. 5.1. Therefore, the particle physicists speak of a "virtual" particle in the intermediate state. However, whatever the terminology, in general a Feynman diagram is not a space-time description of an actual physical process. Of course, the same is true for the right diagram in Fig. 5.2. Here the electron in the initial state at first seems to emit a photon before it meets after a while the incoming photon to absorb it and to turn into the electron in the final state. Hence there are certain limits to the "understanding" of Feynman diagrams. The last instance is always the perturbative calculation of the S-matrix element for the process under consideration and this calculation is completely unambiguous. In particular, the formalism of perturbation theory also predicts that the S-matrix element for Compton scattering in the Born approximation consists of the sum of exactly two amplitudes represented by the two diagrams in Fig. 5.2. Without the explicit calculation, a deeper understanding is hardly possible.

Nowadays, there are computer programs that generate all Feynman diagrams for a process with given initial and final states to a certain order in perturbation theory. These diagrams can then be transformed into graphic files for an eventual publication. Most importantly, the programs also can generate the numerical code for the corresponding S-matrix element to calculate theoretical predictions for comparison with experimental results. Feynman diagrams are indispensable in current particle physics but this was not always the case. Here is an amusing anecdote that is too good not to be true. At the end of the 1940s, the American Nobel Laureate Julian Schwinger developed an alternative approach for QED (Chap. 6) without Feynman diagrams. His students and collaborators at Harvard University were therefore well advised to use the descriptive Feynman diagrams in discussions only with great care. One evening the custodian forgot to lock Schwinger's office after work. The next morning the students passing by were quite astonished after taking a glance at Schwinger's sanctuary. The big blackboard in his office was completely covered with Feynman diagrams. Se non è vero, è ben trovato.

The quantum field theoretic calculation of S-matrix elements guarantees the basic property of quantum physics that amplitudes must be added, not the probabilities. In other words, the probability for Compton scattering in the Born approximation is proportional to the absolute square of the sum of the two amplitudes represented by the Feynman diagrams in Fig. 5.2. The fundamental rule is that for a given initial state all Feynman amplitudes must be added that lead to the same final state. The resulting interferences between the (in general) complex amplitudes are an essential feature of quantum physics and they are of course experimentally accessible.

Before taking a closer look at Compton scattering, we briefly discuss the so-called crossing symmetry of quantum field theory that can be nicely visualised by means of Feynman diagrams. Once the amplitude for Compton scattering $\gamma e^- \to \gamma e^-$ has been calculated, one gets the amplitudes for the processes $\gamma \gamma \to e^- e^+$ (pair creation) and $e^- e^+ \to \gamma \gamma$ (electron-positron annihilation) practically for free. The Feynman diagrams for the latter processes can be obtained by "crossing" some incoming and/or outgoing lines in the diagrams of Fig. 5.2. One must only take into account that by crossing a fermion in the initial state becomes an antifermion in the final state (and vice versa), whereas a photon remains a photon because the photon is its own antiparticle. Then one only has to rename the corresponding energies and momenta and one immediately obtains the amplitudes for the crossed processes.

Back to Compton scattering and to a comparison between theory and experiment. Imagine an incoming beam of photons impinging on a target of electrons, in practice atoms. It is clear that the probability for a scattering event is the bigger the more photons hit the target and the longer the experiment runs. In addition, the probability depends on the number of electrons in the target. To be independent of these parameters that vary from experiment to experiment, the physicists define a so-called cross section for the scattering process under consideration. The cross section is the probability for the scattering on a target particle per time (measured in s^{-1}), divided by the flux of incoming particles, in our case photons (measured in $m^{-2} s^{-1}$). This quotient has obviously the dimension of an area (measured in m^2) and for this reason it is called the cross section for the corresponding scattering (denoted by σ). With some caution, one can therefore interpret the cross section as an effective area of the target particle "seen" by the incoming particle.

To find out what a photon actually sees in Compton scattering, we consider the integrated cross section $\sigma_C = \sigma(\gamma e^- \to \gamma e^-)$. Here one sums for a given initial configuration (in the lab system for an electron at rest defined by the momentum of the incoming photon[5]) over all possible final states. More concretely, one integrates over the momenta and sums over the spins of the particles in the final state. Then the integrated cross section σ_C depends only on the energy E_γ of the incoming photon. We do not carry out the calculation here but we plot the dependence of the cross section[6] on the energy E_γ in Fig. 5.3. More precisely, we plot the ratio $\sigma_C(z)/\sigma_T$ as a function of the ratio $z = E_\gamma/m_e c^2$. Here, σ_T is the so-called Thomson cross section

$$\sigma_T = \frac{8\pi}{3} \left(\frac{\alpha \hbar}{m_e c} \right)^2 = \frac{8\pi}{3} \left(\frac{e^2}{4\pi m_e c^2} \right)^2 . \tag{5.3}$$

Its significance will become clear right away. We have encountered the lengths appearing in this formula in Chap. 3 as classical electron radius r_{cl} and as Compton wave length r_C in Eqs. (3.2) and (3.3). As Fig. 5.3 indicates, the Thomson cross section (5.3) is the low-energy limit of the Compton cross section $\sigma_C(z)$:

$$\sigma_T = \lim_{z \to 0} \sigma_C(z) . \tag{5.4}$$

But σ_T not only is the low-energy limit of the Compton cross section but also its classical limit, i.e. the cross section for the scattering of electromagnetic waves on free electrons in classical electrodynamics. In contrast to the classical case, the Compton cross section is energy dependent and it tends for large energies of the incoming photons ($z \to \infty$) to zero. On the other hand, there are new possible final states in relativistic quantum field theory. For high enough photon energies, electron-positron pairs can be produced in addition to elastic scattering $\gamma e^- \to \gamma e^-$, e.g., $\gamma e^- \to \gamma e^- e^+ e^-$, etc.

The cross section in Fig. 5.3 agrees well with experiment as long as the measurement is not too precise. At the present level of experimental particle physics,

---

[5]We consider unpolarised photons and electrons.
[6]The Compton cross section in lowest-order perturbation theory was first calculated by Oskar Klein and Yoshio Nishina (Klein and Nishina 1929).

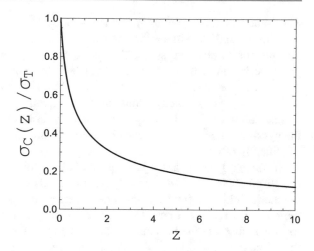

Fig. 5.3 Energy dependence of the ratio $\sigma_C(z)/\sigma_T$ with $z = E_\gamma/m_e c^2$; E_γ is the energy of the incoming photon in the electron rest system

however, the calculation of higher orders in perturbation theory is indispensable. Since Compton scattering is actually a scattering of photons on atoms, the question seems legitimate whether for comparison with a precision experiment one should also take the scattering on the positively charged protons in the atomic nucleus into account. Now the proton is not a fundamental particle but a complicated bound state of quarks and gluons (Chap. 8). Therefore, one cannot simply replace the electron mass by the proton mass in the formula (5.3) for σ_T and in the general formula for $\sigma_C(z)$ to get the cross section for Compton scattering on a proton. But for an order-of-magnitude estimate this is certainly all right. Since the ratio of masses squared of protons and electrons is $m_p^2/m_e^2 \simeq 3 \cdot 10^6$, the contribution of photon-nucleus scattering is indeed negligible.

Anomalous Magnetic Moment of the Electron

At the end of this chapter we turn to another success story of the perturbation theoretic treatment of QED. The magnetic moment of a particle is a measure for the interaction of the particle with a magnetic field. In classical electrodynamics the magnetic moment vector $\vec{\mu}$ for a particle with charge q and mass m is proportional to the orbital angular momentum $\vec{L}$:

$$\vec{\mu} = \frac{q}{2mc}\vec{L} \, . \tag{5.5}$$

In quantum theory there is an additional contribution due to the spin $\vec{S}$ of the particle:

$$\vec{\mu}_S = g\frac{q}{2mc}\vec{S} \, . \tag{5.6}$$

The Landé factor or simply g factor was introduced purely empirically in 1923 by Alfred Landé to explain the anomalous Zeeman effect (Landé 1923). It turned out that for an electron (spin $1/2$) a g factor $g_e \simeq 2$ was necessary to reproduce the experimental results. As a matter of fact, the Dirac equation predicts exactly $g_e = 2$ for the electron. However, after World War II ever more precise measurements showed that g_e is not exactly 2. Since the Dirac equation is a relativistic wave equation, this deviation could not simply be a relativistic correction like the fine structure of the hydrogen atom (Chap. 3). Now it was the turn of quantum field theory and more specifically of QED.

So far we have regarded the field $A_\mu(x)$ in the QED Lagrangian (5.1) as the quantised photon field. But actually $A_\mu(x)$ can also represent a classical field like a magnetic field. The fundamental vertex diagram in Fig. 5.1 can then also be interpreted as the reaction of an electron to a classical magnetic field $\vec{B}$. Thus, it is not too surprising that the Feynman diagram in Fig. 5.1 yields $g_e = 2$, the result from the Dirac equation. But in QED this is only the result in lowest-order perturbation theory. The next-higher order corresponds to the diagram in Fig. 5.4.

Before we discuss the result first calculated by Schwinger, we introduce the convention of particle physicists who call the deviation from the quasi-classical value $g_e = 2$ the anomalous magnetic moment a_e, more precisely $a_e = (g_e - 2)/2$. Since the Feynman diagram in Fig. 5.4 contains two additional vertices compared to the fundamental vertex diagram in Fig. 5.1, a_e must be proportional to e^2, i.e. to α. The Schwinger correction (Schwinger 1948)

$$a_e = \frac{\alpha}{2\pi} \simeq 0.0011614 \tag{5.7}$$

not only was in good agreement with experimental results at the end of the 1940s, but it was also a milestone on the way to general acceptance of QED, in particular of its perturbation theoretic treatment (Chap. 6). In the meantime, 70 years have passed and experimental physics has made tremendous progress. Today, the anomalous magnetic moment of the electron is one of the most precisely measured quantities in physics. The current value is (Hanneke et al. 2011)

Fig. 5.4 Feynman diagram
for the Schwinger correction
to g_e

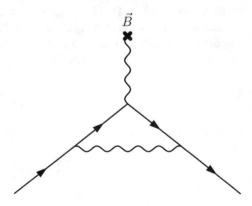

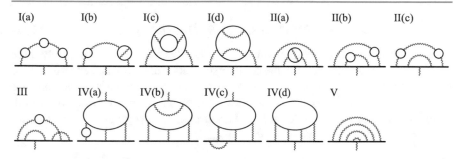

Fig. 5.5 Some of the Feynman diagrams contributing to the anomalous magnetic moment of the electron a_e at order α^4 (From Ayoama et al. 2012; with kind permission of © American Physical Society 2012. All Rights Reserved)

$$a_e^{\text{exp}} = 0.00115965218073(28) \,. \tag{5.8}$$

Comparison with the Schwinger correction (5.7) shows that higher orders in the perturbative expansion are needed if an agreement between theory and experiment should be achieved. Each higher order contributes an extra factor α so that the theoretical result for a_e has the form of a power series in α. For the purpose of illustration, some of the Feynman diagrams needed for the calculation of terms proportional to α^4 are displayed in Fig. 5.5 (Ayoama et al. 2012). All those diagrams have one property in common, namely four independent loops, as one can check with some practice. As a matter of fact, the perturbative expansion in quantum field theory is quite generally an expansion in the number of closed loops in the corresponding Feynman diagrams. We are going to elaborate on this observation in the following chapter.

In QED each closed loop brings in another factor α as can for instance be seen by comparing Figs. 5.4 and 5.5. Today even all five-loop contributions are known (Ayoama et al. 2015). There are exactly 12672 independent Feynman diagrams with five loops! Thus, a_e is known as a power series in the fine-structure constant up to and including order α^5. For comparison with the experimental result (5.8) we therefore need a value for the fine-structure constant. Until recently, the most precise determination from outside particle physics came from precision measurements of the recoil spectrum of rubidium atoms (Bouchendira et al. 2011; Ayoama et al. 2018):

$$\alpha^{-1}(Rb) = 137.035998995(85) \,. \tag{5.9}$$

With this value for α one obtains for the theoretical prediction (Laporta 2017; Ayoama et al. 2018) of the anomalous magnetic moment of the electron

$$a_e^{\text{th}} = 0.001159652182032(720) \,, \tag{5.10}$$

in good agreement with the experimental value (5.8). One can also interpret the result from the other end. Comparing the experimental result (5.8) with the theoretical value for a_e yields the QED value for the fine-structure constant:

$$\alpha^{-1}(\text{QED}) = 137.0359991491(331) \,. \tag{5.11}$$

In comparison with the rubidium value (5.9), the QED value (5.11) is nearly three times as precise. In other words, the comparison between theory and experiment for a_e gave the best value for the fine-structure constant. Very recently, atomic physics has again taken the leadership. A measurement of the recoil frequency of cesium atoms in a matter-wave interferometer (Parker et al. 2018) produced the currently most precise value of the fine-structure constant:

$$\alpha^{-1}(Cs) = 137.035999046(27) \,. \tag{5.12}$$

Putting aside conspiracy theories – after all, the landings on the moon could have happened in a desert in Nevada in reality, the agreement between theory and experiment for the anomalous magnetic moment of the electron is remarkable. With the help of perturbation theory, quantum electrodynamics, which according to some mathematical physicists does not even "exist", makes extremely precise predictions for physical observables that have withstood all experimental tests so far. Since very likely there is no exact solution of QED (see also Chap. 11), the perturbative expansion should not be interpreted as a convergent power series in the fine-structure constant but as a so-called asymptotic series that will be modified by a more fundamental theory at smaller distances. In fact, with the Standard Model (Chap. 9) we already have such an underlying, more fundamental theory at our disposal. Nevertheless, electrodynamics from its classical version as macroscopic Maxwell theory to QED as its quantised version remains the physical theory with the largest domain of validity.

References

Ayoama T et al (2012) Phys Rev Lett 109:111807. https://arxiv.org/abs/1205.5368
Ayoama T et al (2015) Phys Rev D 91:033006. https://arxiv.org/abs/1412.8284
Ayoama T, Kinoshita T, Nio M (2018) Phys Rev D 97:036001. https://arxiv.org/abs/1712.06060
Bouchendira R et al (2011) Phys Rev Lett 106:080801. https://arxiv.org/abs/1012.3627
Hanneke D, Fogwell Hoogerheide S, Gabrielse G (2011) Phys Rev A 83:052122. https://arxiv.org/abs/1009.4831
Heisenberg W (1943) Z Phys 120:513; ibid. 120:673; ibid. 123:93
Klein O, Nishina Y (1929) Z Phys 52:853
Landé A (1923) Z Phys 15:189
Laporta S (2017) Phys Lett B 772:232. https://arxiv.org/abs/1704.06996
Parker RH et al (2018) Science 360:191
Peskin ME, Schroeder DV (1995) An introduction to quantum field theory. Addison-Wesley, Reading
Schwinger J (1948) Phys Rev 73:416
Wheeler JA (1937) Phys Rev 52:1107
Wigner EP (1939) Ann Math 40:149

The Crisis of Quantum Field Theory

6

Infinities of Quantum Field Theory

At the end of the 1920s, quantum electrodynamics in its present form was known. In the following years several scattering processes were calculated in lowest-order perturbation theory. In addition to the reactions related to Compton scattering by crossing (Chap. 5: pair creation $\gamma\gamma \rightarrow e^- e^+$, electron-positron annihilation $e^- e^+ \rightarrow \gamma\gamma$), also elastic electron-electron scattering $e^- e^- \rightarrow e^- e^-$ and elastic electron-positron scattering $e^- e^+ \rightarrow e^- e^+$ were analysed. At that time Feynman diagrams were still unknown and the calculations were considerably more complicated than today. Although QED is Lorentz invariant, a major problem was that the perturbation expansion in the 1930s did not reflect this symmetry manifestly. As a consequence, the contributions of electrons and positrons were treated separately. Especially in higher orders of perturbation theory, this led to some misunderstandings, not to say mistakes.

As far as experimental results were available for the processes mentioned, the agreement with theoretical predictions in the Born approximation was evident. Therefore, initially there were hardly any doubts that QED is the correct theory for the interaction of electrons and positrons with the electromagnetic field. But problems arose as soon as people set out to calculate higher orders in perturbation theory. Let us consider first the fundamental QED vertex diagram in Fig. 5.1. To create a diagram of higher order with the same external lines (particles), we attach an additional vertex on one of the electron lines in the left diagram of Fig. 6.1. That entails an additional factor e but also an additional photon line. Since we want to have a single external photon line only, the additional photon must connect back to an electron line. There are now two possibilities. The extra photon can return to the electron line it originated from or it can connect to the other electron line. We will return to the first possibility shortly. The second possibility leads to the right diagram in Fig. 6.1 depicting the vertex correction that we already encountered in the previous chapter in connection with the anomalous magnetic moment.

© Springer Nature Switzerland AG 2019

G. Ecker, *Particles, Fields, Quanta*, Undergraduate Lecture Notes in Physics,
https://doi.org/10.1007/978-3-030-14479-1_6

Fig. 6.1 Fundamental vertex
diagram of QED (left
diagram) and one-loop vertex
correction (right diagram)

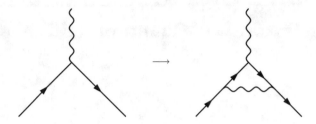

The following properties not only apply to this diagram but to all Feynman diagrams in QED, keeping external lines fixed.

- One order higher in perturbation theory brings with it a factor e^2 in the amplitude. The perturbative series is a power series in the fine-structure constant α.
- One order higher in perturbation theory entails an additional closed loop in the Feynman diagram. The perturbative series is therefore also an expansion in the number of closed loops (see also the four-loop diagrams in Fig. 5.5 generating contributions of order α^4).

QED is a quantum field theory with the elementary charge e as the only coupling constant. Therefore, the loop expansion and the expansion in powers of the fine-structure constant go hand in hand. However, in the Standard Model there are several coupling constants, especially for the part of the weak interactions. How should the perturbative expansion be organised in this general case? There are several good reasons in favour of the loop expansion as the basic ordering principle. First of all, it turns out that the loop expansion corresponds to an expansion in powers of Planck's constant $\hbar$. In contrast to the many different coupling constants in the general case, Planck's constant is a well-defined, experimentally precisely known quantity. As a further strong argument in favour of the loop expansion, the symmetry properties of a quantum field theory are manifest at each order in $\hbar$. For terms with a given power in a coupling constant, this generally is not the case. This is especially important for the gauge symmetries of the Standard Model. The sum of all Feynman amplitudes with the same number of loops (and of course with the same particles in initial and final states) is gauge invariant, an indispensable condition for a meaningful theoretical prediction. In a nutshell, the loop expansion is the measure of all things for the perturbative treatment of any quantum field theory.

Before we turn to the historical development of quantum field theory in the 1930s, we discuss two more ingredients of one-loop Feynman diagrams in QED. As mentioned before in connection with the vertex correction, there is also the possibility of the photon returning to the same electron line it originated from. The corresponding diagram is displayed on the right side of Fig. 6.2 and it is called the self-energy of the electron in the one-loop approximation.

In QED the electron seems to feel an interaction even if there is no other electron or photon anywhere nearby. With the flourishing imagination of the particle physicist, the self-energy correction can be interpreted as an electron that occasionally emits

Fig. 6.2 Free electron (left diagram) and one-loop self-energy correction (right diagram)

Fig. 6.3 Free photon (left diagram) and vacuum polarisation (right diagram)

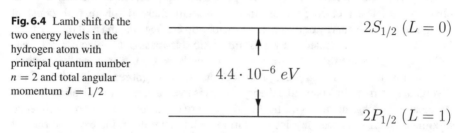

Fig. 6.4 Lamb shift of the two energy levels in the hydrogen atom with principal quantum number $n = 2$ and total angular momentum $J = 1/2$

$2S_{1/2}\ (L = 0)$

$4.4 \cdot 10^{-6}\ eV$

$2P_{1/2}\ (L = 1)$

a (virtual) photon to reabsorb it again immediately afterwards. Similar things may happen to a photon as the right diagram in Fig. 6.3 shows. The photon flies along happily when it suddenly notices that a virtual electron-positron pair wishes to interact with it. It is as if the photon would notice the presence of $e^- e^+$ pairs in the vacuum, hence the name vacuum polarisation. We will discuss very soon what vertex correction, electron self-energy and vacuum polarisation have to do with experimentally measurable effects.

But we will now return to the historical development in the 1930s. In 1930 Robert Oppenheimer, later scientific director of the Manhattan Project in Los Alamos, investigated the influence of QED on the spectral lines of the hydrogen atom. Both the Schrödinger and the Dirac equation predict the degeneracy of some energy levels, i.e., some levels should have the same energy. Oppenheimer rightly conjectured that QED can lift those degeneracies. In particular, he turned to the first excited state of the H-atom with principal quantum number $n = 2$ (Fig. 2.1). According to the Dirac equation the two states with $n = 2$ and total angular momentum $J = 1/2$, which differ in the orbital angular momentum of the electron, have the same energy (Fig. 3.3).

Although in 1930 this had not yet been confirmed experimentally, QED indeed lifts this degeneracy. The splitting is called Lamb shift and it is reproduced in Fig. 6.4. When Oppenheimer tried to calculate the perturbative corrections of the energy levels, he not only discovered that they were different from zero as expected but also that they were actually infinitely large (Oppenheimer 1930)! This was the first

occurrence of the notorious divergences of QED, a seemingly unsolvable problem
for quantum physics during almost two decades.

In 1930 it was already recognized that the divergences of perturbation theory had
to do with the local structure of the interaction. This structure is based on the implicit
assumption that QED would remain valid at smallest distances and therefore also
for arbitrarily high energies. From his unsuccessful attempt Oppenheimer concluded
that QED would make sense at best for energies not exceeding 100 MeV. Today the
world's most powerful accelerator, the LHC at CERN in Geneva, delivers energies
in the order of 10 TeV, a factor 100000 bigger than 100 MeV and QED is still going
strong.

A few years after Oppenheimer, Victor Weisskopf investigated the somewhat
easier problem of the electromagnetic self-energy of the electron (Weisskopf 1934).
Already classical electrodynamics struggled with the problem of the self-mass of
charged particles. Picturing the electron as a small charged sphere with constant
charge density, the energy of the charge distribution changes the mass of the electron.
While this is not unexpected, the electromagnetic self-mass of the electron becomes
infinitely large when the radius of the sphere tends to zero. But since the electron
mass can be measured with great precision, one simply ignores the problem of the
divergent self-mass in Maxwell's theory. There is even a certain justification for this
procedure. Letting the radius of the electron sphere go to zero, one is bound to reach
a domain where classical physics is no more valid. Therefore, the expectation that
quantum field theory and QED in particular would solve this problem was certainly
legitimate. This expectation was bitterly disappointed by the work of Weisskopf
(with help from Wendell Furry). Even though the self-energy of the electron due to
the interaction with the electromagnetic field is "less" divergent than in the classical
case,[1] infinite remains infinite.

The physicists in the 1930s were faced with a dilemma. On the one hand, quantum field theory produced definite successes (spin-statistics theorem, CPT theorem,
agreement between theory and experiment in many cases), but on the other hand
most perturbative corrections to the semi-classical Born approximation were infinite and therefore meaningless. A radical break with the concept of a local quantum
field theory seemed to be called for. Several attempts were undertaken to revoke
strict locality either through modifications at small distances or by limiting the energy range where the theory could be applied. For instance, Heisenberg suggested
that there could be a smallest distance in nature in analogy to the smallest action
$\hbar$ and to the maximal velocity c. All those attempts were unsatisfactory in the end
because even small modifications of the structures not only were arbitrary but they
also had a big impact on measurable quantities. The whole procedure was in conflict
with the well-established idea of the quantum ladder (Chap. 1). From the present
viewpoint of effective quantum field theories (Chap. 11), physics should be able to
describe, e.g., the structure of atoms at least to a good approximation without having

---

[1] The classical self-energy diverges like $1/a$ as the radius a of the sphere goes to zero whereas in
QED the self-energy diverges "only" logarithmically.

to take recourse to new degrees of freedom at highest energies like quantum gravity (Appendix A). The divergences of QED seemed to be in conflict with this well-founded expectation. Therefore, even more radical ideas were put forward, e.g., by Dirac to allow negative probabilities. The situation can be well characterised with a comment from Schwinger (cited in Weinberg 1995): "The preoccupation of the majority of involved physicists was not with analysing and carefully applying the known relativistic theory of coupled electron and electromagnetic fields but with changing it."

The solution of the crisis is a beautiful example for the fruitful collaboration between theory and experiment that has often given rise to significant advances in our understanding of nature. Improvements of experimental methods after World War II led in 1947 to two pioneering precision experiments: the Lamb shift (Lamb and Retherford 1947, Fig. 6.4) and the anomalous magnetic moment of the electron (Kusch and Foley 1947, Chap. 5). Roughly at the same time, a decisive step forward was achieved in the perturbative treatment of QED. In the formulation of Weinberg (1995): "When the revolution came in the late 1940s, it was made by physicists who though mostly young were playing a conservative role, turning away from the search by their predecessors for a radical solution." The theoretical progress was characterised by two essential innovations, the manifestly Lorentz invariant perturbation theory and the concept of renormalisation.

The new formulation of perturbation theory was initiated in 1946 by the Japanese physicist Shinichiro Tomonaga and his collaborators (Tomonaga 1946; Koba et al. 1947; Kanesawa and Tomonaga 1948). In the aftermath of the Second World War, these developments remained unknown in the West for some time. Conversely, Japan was largely cut off from events in the Western world, especially from new experimental results. Allegedly, Tomonaga learned about the experiment of Lamb and Retherford from a short notice in Newsweek. Independently and different from the Japanese approach, Julian Schwinger and Richard Feynman both published their papers on a manifestly Lorentz invariant perturbation theory in 1948 (Schwinger 1948; Feynman 1948). Feynman's article contained the suggestive graphical rules (Feynman diagrams) for the calculation of S-matrix elements. The three approaches (Tomonaga, Schwinger, Feynman) were all different, but Freeman Dyson succeeded to prove (Dyson 1949) that the methods of Tomonaga, Schwinger and Feynman yield the same S-matrix elements. This was one important reason why Feynman diagrams soon prevailed.

Renormalisation

The simplification of the perturbative expansion with manifestly Lorentz invariant methods was by itself not the solution of the divergence problem but it was an important step in the right direction. The solution carries the name renormalisation program of quantum field theory and it not only works for QED but for the Standard Model altogether. The basic idea of renormalisation is a priori independent of perturbation theory and its divergences. Let us return to the starting point of QED, the Lagrangian in Eq. (5.1). That Lagrangian contains two parameters, the mass m and the charge e of the electron, and their interpretation seems to be self-evident. But independently of our intuition, these two quantities must be determined by specific physical measurements.

For the mass this is straightforward. In an experiment energy and momentum of a particle can be measured. By means of the relativistic relation between energy and momentum the mass can then be determined:

$$m = \sqrt{E^2 - p^2c^2}/c^2 \,. \tag{6.1}$$

As discussed above, energy and therefore also the mass are changed by the interaction with the electromagnetic field. The theory must account for this change. In the one-loop approximation this is implemented by the self-energy correction in Fig. 6.2. Therefore, one must distinguish between the parameter in the Lagrangian (for better distinction sometimes denoted as m_0) and the actual physical mass m. In other words, the equality $m = m_0$ only holds at lowest order in perturbation theory.

One could suspect that also the photon becomes massive through the interaction with matter (electrons and positrons in our simplified scenario). Experimental findings (Tanabashi et al. 2018), however, indicate that a possible photon mass would have to be smaller than 10^{-18} eV/c^2. This in turn implies that the Compton wave length of the photon $\lambda_\gamma = \hbar/m_\gamma c$ would be bigger than $2 \cdot 10^{11}$ m, i.e. somewhat longer than the distance between the earth and the sun. This lower bound on λ_γ is not accidental because the best upper limit for m_γ is due to investigations of the magnetic field in the solar wind. In other words, electrodynamics is experimentally validated at least up to distances of $2 \cdot 10^{11}$ m. What does QED have to say? In the one-loop approximation the vacuum polarisation in Fig. 6.3 is the relevant diagram. QED makes a stringent prediction. As long as gauge invariance, i.e. the invariance of QED under the gauge group $U(1)$, is not violated deliberately or accidentally, the photon mass is exactly zero to all orders in perturbation theory.

This brings us to the definition of the electric charge. In this case the definition is not as unique as for the mass, but again gauge invariance guarantees that all methods produce the same value for the charge as long as the measurements are done at low energies. As discussed in the previous chapter, in practice one uses the most precise experimental methods such as measurements of the quantum Hall effect, of the recoil spectra of atoms or of the anomalous magnetic moment of the electron.

For didactic reasons here we use in the discussion of charge renormalisation Compton scattering of photons on electrons that we already analysed in the Born approximation in the previous chapter. There we found that the Compton cross section $\sigma_C(z)$ at lowest order in perturbation theory assumes the form of the Thomson cross section (5.3) in the limit of vanishing photon energy ($z \to 0$):

$$\lim_{z \to 0} \sigma_C(z) = \frac{8\pi}{3} \left(\frac{e^2}{4\pi m c^2} \right)^2 \,. \tag{6.2}$$

The elementary charge e can therefore be defined by the following relation:

$$e = \lim_{z \to 0} \left(\frac{3}{8\pi} (4\pi \, m \, c^2)^2 \sigma_C(z) \right)^{1/4} . \tag{6.3}$$

Since we used the Born approximation with the Feynman diagrams in Fig. 5.2 for the calculation of the Compton cross section, the formulas (6.2) and (6.3) only hold for the parameter in the QED Lagrangian that we should call e_0 in analogy to the mass parameter m_0. If we now turn to the next-higher order in perturbation theory (one-loop order), we first decorate the tree diagrams in Fig. 5.2 with the one-loop insertions vertex correction (Fig. 6.1), self-energy correction (Fig. 6.2) and vacuum polarisation (Fig. 6.3) and then add the remaining one-loop amplitudes. Without further justification, we remark that for the calculation of Feynman amplitudes beyond the Born approximation the field operators also have to be renormalised. At the end of a lengthy calculation one finds that the definition (6.3) now yields a different value for e that is a complicated function of e_0, m_0.

Altogether, at the one-loop level we obtain functions of the form

$$m = m(m_0, e_0), \quad e = e(m_0, e_0) , \tag{6.4}$$

which can be inverted to produce the inverse functions

$$m_0 = m_0(m, e), \quad e_0 = e_0(m, e) . \tag{6.5}$$

With these functions we now express the Feynman amplitudes in the one-loop approximation completely in terms of the physically well-defined parameters m and e. This procedure needs getting used to but it is less complicated than it would seem at a first glance because we only need the Feynman amplitudes to order $\hbar$ relative to the Born approximation. This inversion procedure is always possible iteratively and the renormalisation program then in principle is complete at the one-loop level. All QED amplitudes (not only the one for Compton scattering) and consequently all experimentally accessible quantities like cross sections then only contain the measurable quantities m and e.

Now comes the snag. As they stand, the functions (6.4) are meaningless because the one-loop corrections are infinite. In the renormalisation program sketched above, the divergences of perturbation theory had no reason to evaporate. In the manifestly Lorentz invariant perturbation theory represented by Feynman diagrams, each closed loop comes with a four-dimensional integration over momenta. These integrations are in general divergent because they range from $-\infty$ to $+\infty$ (arbitrarily high energies!). In perturbation theory renormalisation must therefore always be preceded by a so-called regularisation limiting the range of the momentum integration. This may sound almost as arbitrary as the methods of the old perturbation theory in the 1930s. That is true to the extent that many different regularisation procedures can be employed. However, there is an important difference. In the old perturbation theory one just dropped the terms eliminated by regularisation. In contrast, with the new methods of the late 1940s regularisation is always accompanied by renormalisation. In the renormalised amplitudes one can then lift the constraints imposed by the chosen regularisation method. As first shown by Tomonaga and Schwinger in the one-loop approximation, in this way well-defined finite amplitudes, which are independent of the regularisation procedure and which depend only on the physical parameters m and e, are obtained. The method of renormalisation can also be interpreted in the following way. The unknown structure of physics at highest energies (smallest distances) only resides in the renormalised parameters m and e. Consequently, those parameters cannot be calculated in the framework of QED and must be determined experimentally. Soon after Tomonaga and Schwinger, Dyson showed in 1949 that the renormalisation procedure works to all orders of perturbation theory. The article of Dyson also contains a criterion which interaction terms lead to renormalisable quantum field theories. The interaction term of QED in the Lagrangian (5.1) satisfies Dyson's criterion and thus QED belongs to the class of renormalisable quantum field theories. For their fundamental contributions to the perturbative treatment of quantum field theories, Tomonaga, Schwinger and Feynman received the Nobel Prize in 1965. Dyson had the bad luck to come away empty-handed because the Nobel Prize of a given year is attributed to at most three scientists.

The new renormalisation program was immediately applied by several authors (cited in Weinberg 1995, p. 31) to the calculation of the Lamb shift. In one-loop approximation, all (divergent) subdiagrams vertex correction (Fig. 6.1), self-energy correction (Fig. 6.2) and vacuum polarisation (Fig. 6.3) contribute, but the unambiguous final result is finite and in excellent agreement with the experimental value displayed in Fig. 6.4. The successful calculation of the Lamb shift and of the Schwinger correction to the magnetic moment of the electron (Chap. 5) gave QED a tremendous boost. Some even thought that QED had actually been resuscitated by those developments. In their famous textbook on quantum field theory (Bjorken 1965) the authors declared: "QED has achieved a status of peaceful coexistence with its divergences."

In spite of those undeniable successes, the renormalisation program was not accepted by all physicists as solution of the divergence problem of QED. Among them were especially physicists of the older generation like Dirac and Wigner who criticised that "the infinities are only swept under the rug." More surprising was that even Feynman declared as late as 1961 at the occasion of the 12th Solvay Conference (Feynman 1961): "I do not subscribe to the philosophy of renormalisation." Maybe, philosophy was not Feynman's strong point. But also in the Soviet Union the renormalisation program and with it quantum field theory altogether were viewed skeptically in general. The influential physicists Lev Landau and Isaac Pomeranchuk argued on the basis of renormalisation group equations, which will come up again in Chap. 8 in the discussion of quantum chromodynamics, that for asymptotically high energies there would again be divergences. But those energies would be beyond good and evil, in fact much higher than even the Planck energy (Appendix A) where quantum gravity is expected to play a decisive role. Nowadays it is not even possible to frighten students with this so-called Landau pole, often more mysteriously called Landau ghost.

But there was yet another reason for the widespread dissatisfaction with the renormalisation program of QED. Although the heretics had to accept reluctantly the stunning successes of QED perturbation theory, they could rightly refer to the situation of the strong and weak nuclear forces. In the 1920s and 1930s it was recognized that in addition to the electromagnetic interaction also the strong interaction, which holds protons and neutrons together in atomic nuclei, and the weak interaction responsible for instance for β decay had to be taken into account. For the β decay Fermi had actually formulated a quantum field theory (Fermi 1934), but it did not satisfy Dyson's criterion and was therefore not renormalisable. On the other hand, for the strong interaction one could hardly imagine that a perturbative treatment could make sense at all just because of the strength of the nuclear force. Was QED maybe a unique stroke of luck? In the following two chapters we will see how the already given up quantum field theory arose like the famous phoenix from the ashes to serve as the basis of today's Standard Model of all fundamental interactions except gravity.

References

Bjorken JD, Drell SD (1965) Relativistic quantum fields, McGraw-Hill College, New York, p 85

Dyson FJ (1949) Phys Rev 75:486; ibid. 1736

Fermi E (1934) Z Phys 88:161

Feynman RP (1948) Rev Mod Phys 20:367; Phys Rev 74:939; ibid. 1430; ibid. 76:749; ibid. 769

Feynman RP (1961) The quantum theory of fields. In: Proceeding of 12th Solvay conference. Interscience, New York

Kanesawa S, Tomonaga S (1948) Prog Theor Phys 3:1; ibid. 101

Koba Z, Tati T, Tomonaga S (1947) Prog Theor Phys 2:101; ibid. 198

Kusch P, Foley HM (1947) Phys Rev 72:1256; ibid. 73:412; ibid. 74:250

Lamb WE, Retherford RC (1947) Phys Rev 72:241

Oppenheimer JR (1930) Phys Rev 35:461

Schwinger J (1948) Phys Rev 74:1439; ibid. 75:651; ibid. 76:790

Tanabashi M et al (2018) Particle Data Group. Phys Rev D 98:030001

Tomonaga S (1946) Prog Theor Phys 1:27; Phys Rev 74:224

Weinberg S (1995) The quantum theory of fields, vol 1. Cambridge University Press, Cambridge

Weisskopf V (1934) Z Phys 89:27; ibid. 90:817

7

From Beta Decay to Electroweak Gauge Theory

Beta Decay

Of the four fundamental interactions, the weak interaction is in a certain sense the most enigmatic. The two macroscopic interactions, gravitation and electromagnetism, are both familiar to us, at least in the classical version. The strong nuclear force to be treated in the following chapter ensures that nucleons hold together in atomic nuclei. Thus, it is absolutely essential for the existence of matter as we know it. But what do we "need" the weak interaction[1] for? It is of extremely short range and at low energies by far the weakest of the three microscopic interactions. That the weak interaction is responsible for β decay will hardly impress the famous man on the street. But actually also the weak interaction is vital for life on our planet. Without it the production of energy in stars and thus also in the sun would not be possible because the weak interaction plays an essential role in nuclear fusion responsible for energy production. Unlike the sun light, we do not notice the approximately $6.6 \cdot 10^{14}$ neutrinos coming from the sun and hitting the earth per square meter and second. Of course, this is due to the fact that the weak interaction lives up to its name.

Soon after the discovery of radioactivity by Becquerel, Rutherford realised that there must be at least two different sources of radioactivity that he called α and β radiation. In 1900 Becquerel himself carried out the first measurements of the ratio between charge and mass of the β particles. He concluded that the particles were electrons as was confirmed in more precise measurements during the following years. Rutherford and Soddy noticed that radioactivity is not a chemical process but a purely

---

[1] For the nuclear forces physicists use the term "interaction(s)" both in the singular and in the plural, whereas for gravitation and electromagnetism only the singular is common. This may have to do with the fact that in the Standard Model there are several quanta for the strong and weak nuclear forces (eight gluons and $W^{\pm}$, Z, respectively), while there is only one photon and probably only one graviton.

© Springer Nature Switzerland AG 2019

G. Ecker, *Particles, Fields, Quanta*, Undergraduate Lecture Notes in Physics,
https://doi.org/10.1007/978-3-030-14479-1_7

atomic phenomenon. In the predominant picture of the atom at the time, Thomson's plum-pudding model (Chap. 2), one could not specify where the β electrons actually came from. But after the establishment of Rutherford's atomic model the question became more pressing. On the basis of his own atomic model, Bohr concluded that the β electrons could not originate in the atomic shell because their energy was much too high. Thus, β decays must take place in the atomic nucleus. Around 1920, Rutherford realised that the nucleus of the hydrogen atom is contained in all atomic nuclei and he therefore called it proton (Chap. 8). Thus, in an electrically neutral atom there would have to be the same number of protons and electrons. However, if one deduced the number of protons in the nucleus from the number of shell electrons, there was a clear mass deficit for all elements except hydrogen. The total mass of the protons is usually at most half of the atomic mass. Since neutrons were not yet available the physicists concluded that there must be additional protons in the nucleus. Their positive charge would have to be compensated by a corresponding number of electrons in the nucleus. Apparently, those electrons were emitted in β decays.

One could have realised already in 1914 when James Chadwick analysed the energy of β electrons (Chadwick 1914) that something was wrong with that simple model. According to Einstein, the mass difference between the initial and the final nucleus is available (multiplied by c^2, see (Eq. 2.5)) for the energy of the emitted electron. That would mean that for any given initial and final nuclei all electrons emitted in β decays would have the same energy. But in his experiment Chadwick noticed that the electrons were emitted with very different energies, i.e., they had a continuous energy spectrum. As this result appeared to contradict the sacrosanct conservation of energy, it was not generally accepted for several years. But in the 1920s the evidence increased that Chadwick's measurements had been correct after all. In a precision experiment for the β decay $^{210}_{83}Bi \rightarrow \, ^{210}_{84}Po + e^- + ?$, C. D. Ellis and W. A. Wooster found that in spite of a mass difference of $1050\,\text{keV}/c^2$ between the two nuclei the average energy of the decay electrons was only $350\,\text{keV}$ (Ellis and Wooster 1927). At the end of the 1920s Bohr speculated that in the microcosm energy conservation might only hold on average, but that an individual decay could violate the energy balance. But there also was a problem with the conservation of angular momentum if the electron with its spin 1/2 were the only decay product in addition to the final nucleus. In addition, there also were problems with quantum statistics. According to the general picture of the nucleus at the time, the nucleus of the nitrogen isotope $^{14}_{7}N$ should contain 14 protons and 7 electrons. Because of the odd number of particles with spin 1/2 the nitrogen nucleus would have half-integer spin and should therefore satisfy Fermi-Dirac statistics (Chap. 4). But the experiment actually showed that the $^{14}_{7}N$ nucleus had integer spin and was therefore a boson.[2] Finally, it was difficult to reconcile with quantum mechanics and in particular with the uncertainty relation that particles as light as electrons could be confined in such a small volume as an atomic nucleus.

---

[2] The same problem existed for the ^6_3Li nucleus.

Fig. 7.1 Feynman diagram
for the β decay of the
neutron in Fermi theory

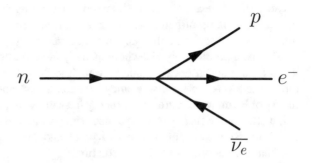

In December 1930 Pauli writes his famous letter to the "Dear radioactive ladies and gentlemen" who gathered for a meeting in Tübingen. He proposes as a solution of the various problems that the electron in β decay is accompanied by an additional particle that would have to be electrically neutral and have spin 1/2. Pauli named the postulated particle neutron but soon the name neutrino (the small neutron) proposed by Fermi was generally accepted. In the presence of this neutrino, the conservation of both energy and angular momentum would be restored. The mass difference between initial and final nuclei determines the total energy of the decay products. This energy is now shared between electron and neutrino, hence the continuous energy spectrum of the electron. Because the neutrino has spin 1/2 the conservation of angular momentum is also guaranteed. Pauli also writes in his letter that he does not dare to publish his idea for the time being. He closes with the regret that he cannot attend the meeting in Tübingen because of a ball in Zürich. Half a year later, at a meeting of the American Physical Society in Pasadena, California in July 1931, Pauli himself presented his neutrino hypothesis but he still prohibited any publication. Only his talk at the 7th Solvay Conference in Brussels in 1933 could finally be published (Pauli 1934).

The discovery of the neutron by Chadwick was the next important step towards understanding β decay (Chadwick 1932). Together with Pauli's neutrino hypothesis, a completely new picture of the atomic nucleus emerged as the starting point of modern nuclear physics. The mysterious bound states of protons and electrons in the nucleus that had been invented to understand atomic masses were no longer needed. Atomic nuclei consist of protons and neutrons only. In β decay electrons and neutrinos are produced in the decay of neutrons:

$$n \rightarrow p + e^- + \overline{\nu_e} . \tag{7.1}$$

On that basis Fermi formulated the first quantum field theory of β decay at the end of 1933, the so-called 4-Fermi theory (Fermi 1934). Although Feynman diagrams did not yet exist, one can characterise the Fermi theory of β decay with the diagram in Fig. 7.1.

The corresponding interaction term does not satisfy Dyson's criterion (Chap. 6) and hence the Fermi theory is a nonrenormalisable quantum field theory. However, because of the weakness of the interaction hardly anybody thought of calculating higher orders in perturbation theory at the time and so the nonrenormalisability was not viewed as a serious deficiency of the Fermi theory.

The main problem lay elsewhere, namely in the structure of the 4-Fermi theory itself. Some nuclear decays could be well described with the Fermi theory while for others there was a clear discrepancy between theory and experiment. The original theory of Fermi was a pure vector theory in analogy to quantum electrodynamics. In a first attempt the theory was expanded with (scalar and tensor) terms, which as the original version were compatible with parity and charge conjugation (Chap. 4). But also that did not turn out to be the solution.

In the meantime new particles had been discovered. In addition to muons, the more massive siblings of the electrons, especially the mesons began to populate the particle zoo. Almost all those particles had also weak decay channels. For instance, muons decay almost exclusively into electrons and two neutrinos[3]:

$$\mu^- \rightarrow e^- + \nu_\mu + \overline{\nu_e} \,. \tag{7.2}$$

Here we have anticipated that six years after the discovery of the electron neutrino (Cowan et al. 1956) also the muon got its own neutrino (Danby et al. 1962). It turned out that in the muon decay (7.2) the strength of the interaction was almost identical with the one governing the β decay of the neutron (7.1), a first indication of the universality of the weak interaction. This universality is a characteristic feature of gauge theories.

Parity Violation and V–A Theory

After the lightest mesons, the pions, also two heavier mesons were found at the end of the 1940s, which apparently decayed into pions due to the weak interaction. The (then) so-called θ meson decayed into two pions ($\theta^+ \rightarrow \pi^+ + \pi^0$), the τ meson into three pions ($\tau^+ \rightarrow \pi^+ + \pi^+ + \pi^-$). Strangely enough, the θ and τ mesons seemed to have practically the same mass and even the same lifetime. This was a strong argument suggesting that the two particles were actually one and the same particle. The (inner) parity of the pion was already known: $P_\pi = -1$, the pions are so-called pseudoscalar particles. If parity were a symmetry of the weak interaction, the parity of the θ meson would then have to be positive: $P_\theta = +1$. On the other hand, from a careful analysis of the energy distribution of the three pions in the decay of the τ meson, the British physicist Richard Dalitz concluded that the parity would have to be negative in this case (Dalitz 1953): $P_\tau = -1$. Then there were only two possibilities: either there was an incredible coincidence and θ and τ were two different particles, or parity is violated by the weak interaction.

---

[3]Since neutrinos are electrically neutral and unaffected by the strong interaction, they are unique messengers of the weak interaction.

Invariance with respect to space reflection had been confirmed in many investigations of atomic and nuclear spectra but this concerned only electromagnetic and strong interactions. Thus, Tsung-Dao Lee and Chen Ning Yang proposed that the question of parity invariance of the weak interaction should be clarified experimentally (Lee and Yang 1956). Until then, most other physicists were convinced that all fundamental interactions were parity invariant. The mere possibility of P violation scandalised Pauli: "Gott ist doch kein schwacher Linkshänder."[4] The experiments proposed by Lee and Yang were carried out without further delay. The results (Wu et al. 1957; Garwin et al. 1957; Friedman and Telegdi 1957) were unambiguous: parity is violated by the weak interaction and the θ-τ puzzle was solved. The two seemingly different particles were indeed one and the same particle that is known as the charged K meson K^+ nowadays.

Although Feynman had just lost a bet because he had backed the wrong horse of parity conservation, he was among the first to draw the right conclusions. Richard Feynman and Murray Gell-Mann and, independently, Robert Marshak and E. C. G. Sudarshan modified the original Fermi theory from a V(ector) to a V(ector)–A(xialvector) theory (Feynman and Gell-Mann 1958; Marshak and Sudarshan 1958). In doing so, they took into account that the measured parity violation was in a certain sense maximal. This can best be seen in the status of neutrinos in the V–A theory. As any particle with spin 1/2, the neutrino has two possible alignments. The spin can either be aligned with the momentum (right-handed neutrino) or opposite to it (left-handed neutrino). However, in the V–A theory only the left-handed neutrino is represented and because of the CPT theorem also the right-handed antineutrino. In this sense, P violation and the asymmetry between left and right are maximal.

This asymmetry can nicely be illustrated with the decay of the charged pion that decays almost exclusively into a muon and its associated neutrino: $\pi^+ \to \mu^+ + \nu_\mu$ and $\pi^- \to \mu^- + \overline{\nu_\mu}$, respectively. Let us consider for definiteness the decay of the negative pion in its rest frame:

$$\pi^-(\vec{P} = \vec{0}) \to \mu^-(-\vec{p}) + \overline{\nu_\mu}(\vec{p}; \text{R}) \ . \tag{7.3}$$

In the rest system of the pion ($\vec{P} = \vec{0}$) the momenta of the muon and of its antineutrino are opposite to each other. The spin of the antineutrino points in the direction of its momentum $\vec{p}$ because according to the V–A theory the antineutrino is right-handed (R). What happens under a parity transformation? The momenta change sign, the spin (as an axial-vector) remains the same and so R(ight) and L(eft) are interchanged. For our pion decay this means:

$$\pi^-(\vec{P} = \vec{0}) \to \mu^-(-\vec{p}) + \overline{\nu_\mu}(\vec{p}; \text{R}) \overset{\text{P}}{\Longrightarrow} \pi^-(\vec{P} = \vec{0}) \to \mu^-(\vec{p}) + \overline{\nu_\mu}(-\vec{p}; \text{L}) \ . \tag{7.4}$$

But in the V–A theory there are no left-handed antineutrinos, i.e., the process on the right side of (7.4) simply does not occur and parity violation is indeed maximal. Let us also have a look at the implication of charge conjugation C, which makes antiparticles

---

[4] After all, God is not a weak left-hander.

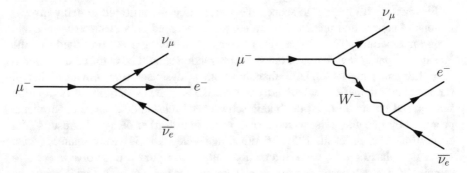

Fig. 7.2 Feynman diagrams for muon decay: V–A theory (left diagram) versus Standard Model (right diagram)

out of particles and vice versa, but leaves momenta and spins untouched. Applying the combined[5] CP transformation to the original decay (7.3) or only C to the right side of (7.4), we obtain the following configuration:

$$\pi^-(\vec{P} = \vec{0}) \rightarrow \mu^-(-\vec{p}) + \overline{\nu_\mu}(\vec{p}; \mathrm{R}) \overset{\mathrm{CP}}{\Longrightarrow} \pi^+(\vec{P} = \vec{0}) \rightarrow \mu^+(\vec{p}) + \nu_\mu(-\vec{p}; \mathrm{L}) . \quad (7.5)$$

The process on the right side of (7.5) is now again a physical process. The positive pion does indeed decay into a μ^+ and a left-handed neutrino, and the probabilities of the decays of π^+ and π^-, i.e. their partial decay rates, are equal. The V–A theory is invariant under CP transformations. Till 1964 nature seemed to respect this prediction, when in decays of the neutral K mesons a small violation of CP symmetry was observed (Christenson et al. 1964). Later on, CP violation was also detected in other decays. We will return to this issue in connection with the Standard Model in Chap. 9.

The V–A theory provides the right description of weak decays, but only at lowest order in perturbation theory (Born approximation). When experimental results became more precise, the question of higher-order corrections became acute. As mentioned above, the Fermi theory and with it the V–A theory are not renormalisable. Therefore, higher-order corrections could not be tackled with the methods known from QED. But what was the deeper reason for this difference? In 1935 the Japanese physicist Hideki Yukawa suspected (Yukawa 1935) that analogous to the photon in QED there would also be quanta of the strong and weak interactions. Although in 1935 only the β decay of the neutron (7.1) was known, we can illustrate Yukawa's suggestion also with the related muon decay (7.2) in Fig. 7.2. Yukawa thereby anticipated an essential aspect of the electroweak gauge theory as it is realised today in the Standard Model. In analogy to the fundamental QED vertex diagram in Fig. 5.1, the four-fermion vertex of the V–A theory is replaced by the vertex diagram in Fig. 7.3.

---

[5]The order of transformations does not matter, CP is as good as PC.

Fig. 7.3 One of the fundamental vertex diagrams of the electroweak gauge theory (ℓ stands for the leptons e, μ and τ); the weak coupling constant g_W takes on the role of the elementary charge e in QED

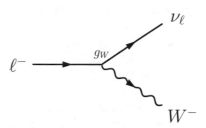

Yukawa also realised that the range of an interaction is related to the mass of the exchanged particle, more precisely to the Compton wave length of the particle. QED is therefore an interaction with infinite range because the photon is massless and that is why we "feel" the electromagnetic interaction in everyday life. In contrast, the weak nuclear force has a very short range and thus the mass M_W of the intermediate vector boson, as it was called originally, must be relatively big. Yukawa also knew that the strength of the weak interaction is governed by the ratio g_W^2/M_W^2, with the weak coupling constant g_W in Fig. 7.3. Therefore, this ratio can for instance be calculated from the muon lifetime.

To deduce the mass of the W boson from the muon lifetime, we need a value for g_W. Since the electroweak gauge theory is addressed in the title of this chapter, we may as well try the hypothesis that g_W is of the same order of magnitude as the elementary charge e. With $g_W \sim e$ the muon lifetime implies $M_W \sim 100\,\text{GeV}/c^2$. Since 1983 we know that our hypothesis was not bad at all.[6] The corresponding Compton wave length $\lambda_W = \hbar/M_W c$ is then $\lambda_W \simeq 2 \cdot 10^{-18}$ m, substantially smaller than even the typical nuclear dimension 10^{-15} m. Incidentally, the structure of the V–A theory also implies that the W boson has spin 1, another similarity with the photon.

The so-called unitarity problem of the V–A theory is related to the nonrenormalisability of the Fermi theory. Let us consider the scattering process

$$\nu_\mu + e^- \rightarrow \mu^- + \nu_e \, , \tag{7.6}$$

related by crossing (Chap. 5) to the muon decay (7.2). Because of the point-like 4-Fermi interaction (left diagram in Fig. 7.2) the cross section $\sigma(\nu_\mu + e^- \rightarrow \mu^- + \nu_e)$ grows quadratically with the center-of-mass energy E. The cross section therefore becomes arbitrarily large for high energies. At some point this must lead to a contradiction with the conservation of probability in quantum field theory. The calculation gives rise to the unitarity bound of the V–A theory $E \lesssim 300\,\text{GeV}$, nearly the same order of magnitude as $M_W c^2$. The proposal of Yukawa solves the unitarity problem. Calculating the scattering amplitude for the process (7.6) with the (crossed version of the) right diagram in Fig. 7.2, the unitarity problem disappears just as in all QED scattering processes.

---

[6]The present value for the mass of the W boson is $M_W = 80.379(12)\,\text{GeV}/c^2$ (Chap. 9).

Electroweak Unification

Before considering a unification of the weak and electromagnetic interactions, several severe obstacles had to be overcome that seemed to stand in the way of such a unification. One such obstacle was the issue of parity, which is a symmetry of QED but is even maximally violated by the weak interaction. As the example of the neutrinos shows, gauge bosons (in this case W bosons) can interact differently with left- and right-handed components of fermion fields.[7] This also applies to charged fermions, both to leptons and quarks. Somewhat exaggerating, parity violation is actually the normal situation in gauge theories. Only if the interactions of the corresponding gauge bosons do not distinguish between left- and right-handedness, parity can be conserved. This is the case in both QED and QCD (Chap. 8).

W bosons are electrically charged and are therefore also subject to the electromagnetic interaction. What is the form of these interactions in a unified electroweak quantum field theory? The following argument does not correspond to the actual historical development, but was put forward only a posteriori. Nevertheless, it is a beautiful example of enlightening theoretical deliberations (Llewellyn Smith 1973). Requiring that not only the scattering $\nu_\mu + e^- \rightarrow \mu^- + \nu_e$ is compatible with unitarity, but all scattering processes with fermions (leptons and quarks) and vector bosons (photon, $W^\pm$), the underlying quantum field theory must have the structure of a so-called non-abelian gauge theory. Such theories were first formulated by C. N. Yang and Robert L. Mills (Yang and Mills 1954). They are a generalisation of QED with the abelian gauge group $U(1)$ and the photon as single gauge boson (Chap. 5) to a quantum field theory with a (more complicated) non-abelian gauge group and several gauge bosons, which can interact with each other as in the case of photon and W boson.

In the original version of their theory, Yang and Mills actually had in mind the strong interactions of mesons (pions, kaons, vector mesons) but this idea was soon put aside. The main reason was that gauge invariance would require massless vector mesons, which still do not exist today. For an application of Yang-Mills theories to electroweak unification, this circumstance once again seemed to stand in the way. Indeed, the photon is massless as a decent gauge boson should be, but this is certainly not the case for the W boson, which is after all responsible for the short-range weak interaction. It seemed that theory had again reached an impasse. A unitary, renormalisable quantum field theory of electroweak interactions requires the invariance of the theory under a (non-abelian) gauge symmetry, which seemed to imply the existence of massless W bosons, in striking contradiction with the structure of the weak interaction.

The way out of this dilemma was due to the recognition that in quantum field theory, unlike in classical physics, there are two different ways how a symmetry

---

[7]For massive fermions, theoreticians use the notion of chirality instead of handedness for reasons of Lorentz invariance. Strictly speaking, the two concepts agree only for massless fermions. Nevertheless, here we use for simplicity for all fermions, including massive ones, the more descriptive handedness.

Fig. 7.4 Schematic
representation of spin
alignment in Weiss domains
of a ferromagnet

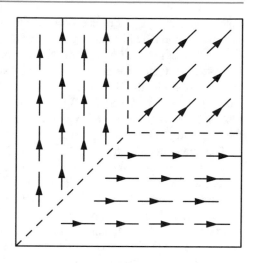

can manifest itself. In both manifestations the field equations and the Lagrangian remain unchanged under symmetry transformations. How the state of lowest energy (ground state) behaves under symmetry transformations, is responsible for the difference between the two variants of symmetry implementation in quantum field theory. The physics of the ferromagnet is an instructive example. In the absence of an external magnetic field, the underlying theory of the ferromagnet does not single out a special direction, the theory is rotationally invariant. Nevertheless, below a critical temperature and within certain domains, the ground state is characterised by a certain direction along which the spins are aligned (Weiss domains, Fig. 7.4). Because of the rotational invariance of the underlying equations, each direction is equally good, it is apparently chosen "spontaneously". This has led to the notion of spontaneous symmetry breaking. The name is a bit misleading because the symmetry is not really broken. After all, the field equations and the Lagrangian are still invariant. This fact will be important for the realisation of gauge symmetry in the electroweak gauge theory.

In particle physics the phenomenon of spontaneous symmetry breaking was first investigated by Yoichiro Nambu who had in mind a realisation in the context of the strong interaction (Nambu 1960). Motivated by Nambu, Jeffrey Goldstone showed in the following year that in general massless particles (Nambu-Goldstone bosons) occur together with spontaneous symmetry breaking (Goldstone 1961). Incidentally, this explains why pions are the lightest hadrons (strongly interacting particles). But for electroweak unification the Goldstone theorem seems to be counterproductive. We do not need any massless bosons but an explanation for the mass difference between photon and W boson. This explanation was provided by Robert Brout and François Englert (Englert and Brout 1964) and, independently, by Peter Higgs (Higgs 1964). While in theories with spontaneously broken global symmetries massless particles do indeed occur, the Nambu-Goldstone bosons of a gauge symmetry (local symmetry) are not physical particles but they can instead bestow a mass upon some of the originally massless gauge bosons. Without the appropriate mathematical background, this may sound a bit like hocus-pocus but the counting of degrees of freedom makes the mechanism at least more plausible. As mentioned earlier, all massless particles with spin $\neq 0$ like the photon only have two degrees of freedom. But a massive vector boson like the W boson (spin 1) has three degrees of freedom – in general $2S + 1$ states for spin S. The B(rout)-E(nglert)-H(iggs) mechanism precisely furnishes the missing degree of freedom (also called "would-be-Goldstone boson" in the literature) to make a massive W boson out of a massless one.

The phenomenon actually also occurs in the physics of condensed matter, namely in superconductors (Anderson 1958). In certain metals so-called Cooper pairs can form, i.e. nonlocal bound states of two electrons with opposite momenta and spin directions. Effectively, those bound states are then bosons with charge $-2e$ that can condense in the ground state at sufficiently low temperatures (Bose-Einstein condensation, Chap. 4). This ground state has therefore negative charge, which breaks the $U(1)$ invariance of QED spontaneously. But where are the massless Nambu-Goldstone bosons of superconductivity? The answer is known as Meißner-Ochsenfeld effect. An applied magnetic field cannot propagate freely in a superconductor but it can only penetrate up to a certain penetration depth. In the language of quantum field theory, this penetration depth corresponds to a finite Compton wave length and therefore to an apparent mass – of course only in the superconductor, outside the photon is again massless as it should be. Since QED is a gauge theory with the gauge group $U(1)$, also in a superconductor the Nambu-Goldstone boson is "eaten up" by the photon to acquire an apparent mass.

Thus all necessary preconditions for a consistent quantum field theory of electroweak interactions were in place. The renormalisability of such a theory was proved by the Dutch physicists Gerard 't Hooft and Martinus Veltman at the beginning of the 1970s ('t Hooft 1971; 't Hooft and Veltman 1972). This was the start of a fruitful collaboration between theory and experiment where ever more precise experiments confirmed ever more detailed theoretical predictions. A first highlight of this period lasting until today was the detection of the predicted gauge bosons $W^{\pm}$, Z in experiments at CERN at the beginning of the 1980s (for more details, see Chap. 9).

References

Anderson PW (1958) Phys Rev 112:1900
Chadwick J (1914) Verhandlungen der Deutschen Phys Ges 16:383
Chadwick J (1932) Nature 129:312
Christenson JH, Cronin JW, Fitch VL, Turlay R (1964) Phys Rev Lett 13:138
Cowan CL et al (1956) Science 124:103
Dalitz RH (1953) Phil Mag 44:1068; Phys Rev 94:1046
Danby G et al (1962) Phys Rev Lett 9:36
Ellis CD, Wooster WA (1927) Proc Royal Soc A 117:109
Englert F, Brout R (1964) Phys Rev Lett 13:321

Fermi E (1934) Z Phys 88:161
Feynman RP, Gell-Mann M (1958) Phys Rev 109:193
Friedman JI, Telegdi VL (1957) Phys Rev 105:1681
Garwin RL, Lederman LM, Weinrich M (1957) Phys Rev 105:1415
Goldstone J (1961) Nuovo Cim 19:154
Higgs PW (1964) Phys Rev Lett 13:508
't Hooft G (1971) Nucl Phys B 35:167
't Hooft G, Veltman M (1972) Nucl Phys B44:189; ibid. B50:318
Lee TD, Yang CN (1956) Phys Rev 104:254
Llewellyn Smith CH (1973) Phys Lett B 46:233
Marshak RE, Sudarshan ECG (1958) Phys Rev 109:1860
Nambu Y (1960) Phys Rev 117:648
Pauli W (1934) Rapports du Septième Conseil de Physique Solvay, Bruxelles 1933. Gauthiers-
     Villars, Paris
Wu CS et al (1957) Phys Rev 105:1413
Yang CN, Mills RL (1954) Phys Rev 96:191
Yukawa H (1935) Proc Phys Math Soc Japan 17:48

Quantum Chromodynamics: Quantum Field Theory of the Strong Interaction

8

Strong Nuclear Force

As early as 1815, the English chemist William Prout suspected on the basis of existing measurements of atomic masses that all atoms are built up of hydrogen atoms (Prout's hypothesis). In tribute to him, Rutherford proposed in 1920 at a meeting of the British Association for the Advancement of Science that the nucleus of the hydrogen atom should be called prouton or proton. Apparently, proton was more popular.

A few years earlier, Rutherford had recognized that the hydrogen nucleus does indeed occur in all nuclei but that additional, electrically neutral constituents must be contained in the nuclei in order to understand nuclear masses. He called those constituents neutrons and pictured them as bound states of protons and electrons. Two reasons seemed to support such a picture. The mass of the neutrons was comparable with the proton mass and the negatively charged electrons would compensate the electrostatic repulsion of the protons at least to some extent. Scattering α particles on protons, Rutherford had noticed deviations from the Coulomb law (electrostatic repulsion between α and p) but he associated those deviations with the complex nature of the α particles. The first published indication for the existence of an additional "strong nuclear force" is due to Rutherford's assistant James Chadwick and his collaborator Etienne Bieler (Chadwick and Bieler 1921). They also had investigated the scattering of α particles on protons. For slow α particles, the angular distribution of the recoil protons was as expected (Rutherford scattering) but for more energetic α particles clear deviations showed up: "It is our task to find some field of force which will reproduce these effects."

But that was easier said than done. As discussed at the beginning of the previous chapter in connection with β decay, during the 1920s more and more problems turned up for the popular model of the atomic nucleus consisting of protons and electrons (nuclear spins, uncertainty relation, etc.). All those problems evaporated after the discovery of the neutron by Chadwick (1932). The atomic nucleus consists

© Springer Nature Switzerland AG 2019

G. Ecker, *Particles, Fields, Quanta*, Undergraduate Lecture Notes in Physics,
https://doi.org/10.1007/978-3-030-14479-1_8

of protons and neutrons only – neither electrons nor neutrinos in the nucleus. Thus, there must exist a "strong nuclear force" preventing protons from flying apart because of the electrostatic repulsion. In the same year Heisenberg introduced the concept of isospin symmetry of the strong interaction to explain the approximate equality of proton and neutron masses. It is important to realise that isospin symmetry not only explains the near equality of masses but also the approximate equality of proton-proton, proton-neutron and neutron-neutron forces. During the following years, those predictions were confirmed in investigations of nuclear spectra.

But what is it that keeps protons and neutrons together in atomic nuclei? In the same article where Yukawa had postulated a carrier for the weak nuclear force (to-day's W boson, Chap. 7), he also suggested that the strong nuclear force could be generated by the exchange of a hypothetical particle with spin 0. The mass of this particle would have to be about 200 times bigger than the electron mass and that is why Yukawa's particle was originally called mesotron (between electron and proton). In the course of time, the mesotron lost three letters and turned into a meson, following a suggestion of Heisenberg. Already in 1937 Carl Anderson and Seth Neddermeyer detected a particle with approximately the corresponding mass in cosmic rays (Anderson and Neddermeyer 1937). But this particle hardly interacted with nuclear matter and therefore could not be Yukawa's meson. Anderson and Neddermeyer had actually detected muons, the more massive siblings of electrons. But ten years later, Powell et al. observed charged pions ($\pi^{\pm}$ mesons) in cosmic rays using photoemulsions (Lattes et al. 1947). From that date on, the physics of mesons has been an integral part of nuclear physics. On the basis of nonrelativistic potential models involving mesons, significant progress was made in the understanding of the forces between nucleons. Those potentials, attractive at long distances (pion exchange) and strongly repulsive at short distances, provided satisfactory explanations of nuclear structure and of nuclear reactions. In particular, it became possible to understand the mechanisms for the production of energy in stars and therefore the luminosity of the sun (Bethe 1938; von Weizsäcker 1937).

However, those significant achievements were restricted to reactions where nucleons had small relative velocities. For processes with relativistic nucleons, nuclear potentials were inadequate just as one cannot describe Compton scattering with the Coulomb potential. In principle, there was no problem in constructing quantum field theories with nucleons and pions that were even renormalisable. But it is no accident that the strong interaction carries its name. Perturbation theory, so successful for electromagnetic and weak interactions, was simply not applicable because of the strength of the strong interaction.

In spite of the undeniable successes of quantum field theory (Chaps. 4, 5, 6 and 7), several physicists therefore speculated that the strong interaction might not be described with a local quantum field theory. Here are two opinions of prominent skeptics.

- Landau: "It is well known that theoretical physics is at present almost helpless in dealing with the problem of strong interactions. We are driven to the conclusion that the Hamiltonian method for strong interactions is dead and must be buried, although of course with deserved honour." (Landau 1960)

- Marvin Goldberger: "My own feeling is that we have learned a great deal from field theory ...that I am quite happy to discard it as an old, but rather friendly, mistress who I would be willing to recognize on the street if I should encounter her again." (Goldberger 1961)

In the 1960s people were therefore looking for alternatives to quantum field theory to get a handle on the strong interaction. At that time, more and more strongly interacting particles (hadrons) were found, which could not possibly all be fundamental. This impression prompted the launching of "nuclear democracy": all hadrons are equal and it therefore makes no sense to set up a quantum field theory with only a few of them as fundamental fields. At that time, this idea was politically absolutely correct but the implementation, called S-matrix theory or bootstrap method (Chew 1962), suffered the same fate as the student movement of the late 1960s. The great expectations could not be fulfilled. The restriction to general properties of the S-matrix, which were expected to lead to measurable predictions by means of self-consistency conditions (bootstrap!), was not effective in the end. A severe blow to S-matrix theory was the realisation that the elastic scattering of pions becomes weaker at low energies. This fact was in striking contradiction to basic assumptions of S-matrix theory (Weinberg 1999).

From the Quark Model to Asymptotic Freedom

A different approach to the physics of hadrons was advocated especially by Gell-Mann. His strategy consisted in extracting certain algebraic relations from quantum field theoretic models, which were then no longer taken seriously. Gell-Mann described this "recipe" in the following way (Gell-Mann 1964a): "We construct a mathematical theory of the strongly interacting particles, which may or may not have anything to do with reality, find suitable algebraic relations that hold in the model, postulate their validity, and then throw away the model. We may compare this process to a method sometimes employed in French cuisine: a piece of pheasant meat is cooked between two slices of veal, which are then discarded."

This recipe was quite successful, in particular for the formulation of the quark model (Gell-Mann 1964b; Zweig 1964). All hadrons (mesons and baryons) known at the time could be understood as bound states of three fictitious constituents (quarks u, d, s) with peculiar electric charges (Chap. 9). For instance, the proton consists of two u quarks and one d quark ($p \sim uud$), while the negatively charged pion corresponds to a state with one anti-u quark and one d quark ($\pi^- \sim \bar{u}d$). Strangely enough, these quarks could not be isolated experimentally. The underlying dynamics of the quark model remained a mystery. In the spirit of his culinary philosophy, Gell-Mann insisted till the beginning of the 1970s that the quarks were purely hypothetical quantities ("mathematical entities") without physical reality. Another success story

Fig. 8.1 Schematic representation of deep inelastic electron-nucleon scattering $e^- \, N \rightarrow e^- \, X$. In the cross section one sums over all possible hadronic final states X where the total charge of X equals the nucleon charge

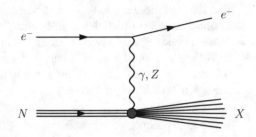

was current algebra that we understand today as a consequence of the approximate chiral symmetry of quantum chromodynamics. The previously mentioned analysis of pion-pion scattering at low energies was conducted in the framework of current algebra. In spite of these successes deeper insights into the dynamics of the strong interaction were lacking.

As it is often the case, experiment (SLAC–MIT, Breidenbach et al. 1969) provided the decisive hints. In the so-called deep inelastic scattering of leptons (electrons, muons, neutrinos) on nucleons (Fig. 8.1), the cross sections displayed an increasingly simple structure at high energies and for large transverse momenta (Nobel Prize 1990 for Jerome Friedman, Henry Kendall and Richard Taylor). Under those conditions, the nucleons appear as collections of free particles, which were called partons by Feynman. The quarks of Gell-Mann and Zweig were obvious candidates for those partons. But this generated a seemingly intractable dilemma for a quantum field theory of the strong interaction. How could the quarks be so strongly bound in mesons and baryons that they could not be isolated as free particles, if on the other hand they behaved as quasi-free particles in deep inelastic scattering? This seemed to be yet another argument in favour of the view that the quarks were purely mathematical constructs without physical reality.

That the strength of an interaction may depend on energy was actually a well-known phenomenon in QED. An illustrative explanation of the so-called charge screening is shown in Fig. 8.2. In QED there are vacuum fluctuations of virtual electron-positron pairs (Chap. 5) that screen the real charges. The greater the distance between two real charges (electrons in Fig. 8.2), the more they are screened by the virtual e^+e^- pairs. Therefore, the effective charges responsible for the strength of the interaction decrease with increasing distance. Conversely, the effective charges increase the closer they come. In quantum field theory smaller distances correspond to higher energies and vice versa. QED is therefore denoted as ultraviolet unstable because the strength of the electromagnetic interaction gets larger with increasing energy (decreasing distance). Incidentally, this is the reason for the Landau pole in QED mentioned at the end of Chap. 6.

The phenomenon of charge screening is not restricted to QED. Theoreticians investigated various renormalisable quantum field theories with regard to their behaviour at high energies. The aim was to find a theory where the high-energy behaviour is different from the one in QED because the results of deep inelastic scattering pointed to an ultraviolet stability of the strong interaction. In other words, one

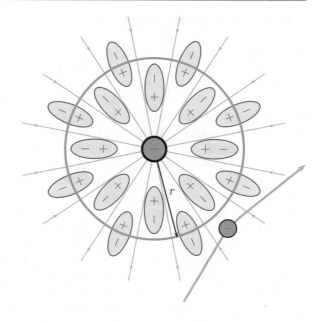

was looking for an "asymptotically free" quantum field theory where the interaction becomes weaker at higher energies. This search reached its climax in 1973. At the beginning of that year a seeming setback happened when Tony Zee showed that charge screening and therefore ultraviolet instability occur in a large class of quantum field theories (Zee 1973). In the abstract of his paper he writes: "On the basis of this result we conjecture that there are no asymptotically free quantum field theories in four dimensions." At this time, Sidney Coleman and David Gross worked on the same problem (Coleman and Gross 1973), with the declared purpose to show that a renormalisable quantum field theory cannot be asymptotically free. By that time both theoretical and experimental results indicated that the electromagnetic and weak interactions can be described by a unified Yang-Mills theory (Chap. 7). Although Yang and Mills had originally planned to apply their theory to the strong interaction, that option was soon abandoned for various good reasons. To close the "loophole" in the presumed incompatibility between renormalisable quantum field theories and asymptotic freedom, the graduate students David Politzer (Harvard, supervisor Coleman) and Frank Wilczek (Princeton, supervisor Gross) were assigned to investigate the high-energy structure of non-abelian gauge theories. The outcome is part of the history of physics (Nobel Prize 2004 for Gross, Politzer and Wilczek). In two successive papers in the Physical Review Letters (Gross and Wilczek 1973; Politzer 1973), the asymptotic freedom of Yang-Mills theories was proved as long as there are not too many fermions in the theory (see below). The way was open for a non-abelian gauge theory of quarks and gluons later to be called quantum chromodynamics (QCD). Gross: "For me the discovery of asymptotic freedom was completely unexpected. Like an atheist who has just received a message from a burning bush, I became an immediate true believer." (Gross 1999)

Quark-Gluon Gauge Theory

At that time it was already known that each quark flavour $u, d, s, \ldots$ possesses an additional degree of freedom called "colour". Quarks come in three different colours and each particle physicist may choose her (his) own favourite colours. Of course, this colour has absolutely nothing to do with the optical colours. Colour is not a fantasy of underemployed particle physicists but is supported by several experimental indications. On the other hand, it was not clear before 1973 how many gluons there are as carriers of the strong interactions. For group theoretic reasons there are only two possibilities assuming that the quarks come indeed in three colours. Either there is only one gluon just as there is only one photon in QED or the strong interactions are mediated by eight gluons. Only in the latter case we have a non-abelian gauge theory and, consequently, asymptotic freedom only holds in that case. In an influential paper, Harald Fritzsch, Murray Gell-Mann and Heinrich Leutwyler presented five arguments for an octet of gluons in the fall of 1973 (Fritzsch et al. 1973). Argument number four was the asymptotic freedom of non-abelian gauge theories that had been discovered a few months earlier. For good reason, the year 1973 is therefore known as the year of birth of QCD.

Before we take a closer look at asymptotic freedom, we are going to confront the Lagrangian of QED for the electron with the Lagrangian of QCD for one quark flavour. To start with, we write the QED Lagrangian (5.1) in a more compact form as

$$\mathcal{L}_{\text{QED}} = \overline{\psi} \left(i \not{D} - m_e \right) \psi - \frac{1}{4} F_{\mu\nu} F^{\mu\nu} . \tag{8.1}$$

Here, $\not{D}$ stands for $\gamma^\mu D_\mu$, where $D_\mu = \partial_\mu - i\, e\, A_\mu$ is a so-called covariant derivative. The QCD Lagrangian has a similar form:

$$\mathcal{L}_{\text{QCD}} = \sum_{i,j=1}^{3} \overline{q_i} \left(i \not{D}_{ij} - m_q \delta_{ij} \right) q_j - \frac{1}{4} \sum_{\alpha=1}^{8} G_{\mu\nu}^\alpha G^{\alpha,\mu\nu} . \tag{8.2}$$

The electron field ψ in Eq. (8.1) is replaced by three quark fields q_i ($i = 1, 2, 3$ denote the three colours) of a given quark flavour. The covariant derivative $D_{\mu,ij}$ is now a 3×3 matrix in colour space. The photon field A_μ is substituted by eight gluon fields G_μ^α ($\alpha = 1, \ldots, 8$),

$$D_{\mu,ij} = \delta_{ij} \partial_\mu + \frac{i}{2} g_s (\lambda_\alpha)_{ij} G_\mu^\alpha , \tag{8.3}$$

multiplied by the so-called Gell-Mann matrices[1] λ_α. The Kronecker symbol δ_{ij} is an explicit representation of the unit matrix in 3-dimensional colour space ($\delta_{ij} = 1$ for

---

[1]The 3×3 matrices λ_α form a 3-dimensional representation of the Lie algebra of $SU(3)$ in a basis suitable for particle physics. This Lie algebra is characterised by the structure constants $f_{\alpha\beta\gamma}$ ($\alpha, \beta, \gamma = 1, \ldots, 8$) in (8.4).

Fig. 8.3 Fundamental
quark-gluon vertex diagram
in QCD (without colour
indices)

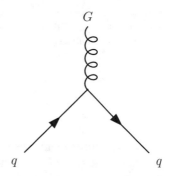

$i = j$, $\delta_{ij} = 0$ for $i \neq j$). In the Lagrangian (8.2) the Kronecker symbol indicates
that the three colours of a given quark flavour all have the same mass m_q. Gauge
invariance requires a single coupling constant g_s, i.e., all eight gluons couple with
the same strength to the quarks (universality, see also section "Extensions of the
Standard Model" in Chap. 10).

The formal similarity of the Lagrangians (8.1) and (8.2) is due to the fact that
both theories are gauge theories.[2] But QED is an abelian gauge theory (ultraviolet
unstable) with the gauge group $U(1)$, whereas QCD is a non-abelian gauge theory
(ultraviolet stable = asymptotically free) with the gauge group $SU(3)$. In particular,
this can be seen in the difference between the electromagnetic field strength tensor
$F_{\mu\nu}$ and the gluonic field strength tensors $G^\alpha_{\mu\nu}$ ($\alpha = 1, \ldots, 8$):

$$G^\alpha_{\mu\nu} = \partial_\mu G^\alpha_\nu - \partial_\nu G^\alpha_\mu - g_s f_{\alpha\beta\gamma} G^\beta_\mu G^\gamma_\nu \, . \tag{8.4}$$

While in an abelian gauge theory like QED the field strength tensor $F_{\mu\nu}$ is linear in
the gauge field (Chap. 5), gauge invariance in a non-abelian gauge theory requires
also terms quadratic in the gauge fields in the field strength tensors (8.4), multiplied
by the gauge-coupling constant g_s.

The difference between QED and QCD also manifests itself in the fundamental vertices of the two theories. In QED there is only one electron-photon vertex
(Fig. 5.1). The analogue in QCD is the quark-gluon vertex in Fig. 8.3. Because both
gluons and quarks carry colour charge, in QCD there are two more fundamental
vertices (absent in QED because the photon is electrically neutral) that appear in
the second term of the Lagrangian (8.2) describing the self-interaction of gluons
(Fig. 8.4). As can be seen from the form of the gluonic field strength tensors (8.4),
these vertices correspond to terms in the QCD Lagrangian (8.2) that are cubic or quar-
tic in the gluon fields and linear or quadratic, respectively, in the coupling constant
g_s.

But why is QCD asymptotically free, in contrast to QED? With the current tools
of theoretical particle physics, this question can be answered in one afternoon in all

---

[2] M. Gell-Mann: "It's all symmetries!", private communication at a Viennese Heurigen, Sept. 2011.

Fig. 8.4 3- and 4-gluon
vertex diagrams of QCD
(without colour indices)

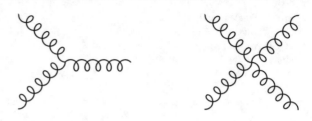

details. It suffices to investigate the divergence structure of a few one-loop diagrams
to deduce the result by means of renormalisation group equations. It can be seen
as an irony in the history of physics that in a certain sense asymptotic freedom was
always hidden under the rug under which the divergences had supposedly been swept
(Chap. 6).

Interpreting the vacuum of a quantum field theory as a polarisable medium
(Nielsen 1981), one can obtain a physical understanding of asymptotic freedom.
Unlike in a usual macroscopic medium, the product of permittivity ε and perme-
ability μ is always equal to one in a relativistic theory. A "normal" quantum field
theory like QED has $\varepsilon > 1$ because of charge screening and so the vacuum can also
be viewed as a (colour) diamagnet ($\mu < 1$). In a non-abelian gauge theory like QCD
not only fermions (quarks in our case) carry a (colour) charge but also the quanta
of the interactions, in our case the "coloured" gluons. The gluons, which have spin
1 (gauge theory!) like the photon, act therefore like permanent magnetic dipoles
making the vacuum paramagnetic ($\mu > 1$). As a matter of fact, the charge screening
of the quarks and colour paramagnetism of the gluons act in different directions.
For the asymptotic freedom of a gauge theory with gauge group $SU(N_c)$ the sign
of the expression $2 N_f - 11 N_c$ matters where N_f is the number of quark flavours.
The gauge theory in question is asymptotically free if and only if $2 N_f - 11 N_c$ is
negative. Therefore, with three colours ($N_c = 3$) QCD is asymptotically free as long
as there are not more than 16 quark flavours. As nature seems to be getting along
with six quark flavours u, d, c, s, t, b (Chap. 9), QCD is asymptotically free.

The message of the "burning bush" was not immediately acknowledged by all
particle physicists. But in the course of time, experimental indications accumulated
that the strength of the strong interactions actually depends on energy. In Fig. 8.5
the decrease of the coupling strength α_s (in analogy to QED denoted as strong fine-
structure constant[3]) with energy is displayed. Over three orders of magnitude in the
energy, the theoretical prediction of QCD agrees with the various experimental results. The strong fine-structure constant α_s is therefore known today with an accuracy
of 1%.

In the flowery language of particle physicists, the reverse side of asymptotic
freedom is sometimes called "infrared slavery". For low energies (large distances),

---

[3] Since the discovery of asymptotic freedom at the latest, particle physicists are aware that coupling
constants are not constants in the usual sense but depend on energy. Nevertheless, the names coupling
constant, fine-structure constant, etc. are still being used.

Fig. 8.5 Dependence of the strong fine-structure constant $\alpha_s = g_s^2/(4\pi\hbar c)$ on the energy Q (in GeV). Note the logarithmic scale for the energy (From Tanabashi et al. 2018; with kind permission of © Particle Data Group 2018. All Rights Reserved)

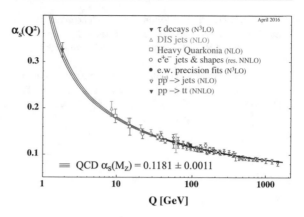

the effective coupling increases as Fig. 8.5 shows. Many theoretical arguments and especially the experimental situation suggest that quarks and gluons are permanently confined in hadrons (confinement). In contrast to asymptotic freedom, confinement has so far not been deduced from the underlying equations of QCD (based on the Lagrangian (8.2)). This also has to do with the problem that perturbation theory is not applicable for large coupling strength.

References

Anderson CD, Neddermeyer SH (1937) Phys Rev 51:884

Bethe HA (1938) Phys Rev 55:434

Breidenbach M et al (1969) SLAC-MIT-experiment. Phys Rev Lett 23:935

Chadwick J, Bieler ES (1921) Phil Mag 42:923

Chadwick J (1932) Nature 129:312

Chew GF (1962) Rev Mod Phys 34:394

Coleman S, Gross DJ (1973) Phys Rev Lett 31:851

Fritzsch H, Gell-Mann M, Leutwyler H (1973) Phys Lett B 47:365

Gell-Mann M (1964a) Physics 1:63

Gell-Mann M (1964b) Phys Lett 8:214

Goldberger M (1961) The quantum theory of fields. In: Proceeding of 12th Solvay conference. Interscience, New York

Gross DJ (1999) Nucl Phys Proc Suppl 74:426. https://arxiv.org/abs/hep-th/9809060

Gross DJ, Wilczek F (1973) Phys Rev Lett 30:1343

Jegerlehner F (2008) The anomalous magnetic moment of the muon. Springer Tracts Mod Phys 226. http://dx.doi.org/10.1007/978-3-540-72634-0

Landau LD (1960) Fundamental problems. Pauli Memorial Volume, Interscience, New York

Lattes CMG, Muirhead H, Occhialini GPS, Powell CF (1947) Nature 159:694

Nielsen NK (1981) Am J Phys 49:1171

Politzer HD (1973) Phys Rev Lett 30:1346

Tanabashi M et al (2018) Particle Data Group. Phys Rev D 98:030001

Weinberg S (1999) Proc. of Symposium on Conceptual Foundations of Quantum Field Theory, 241, Ed Cao TY. Cambridge Univ. Press, Cambridge https://arxiv.org/abs/hep-th/9702027
von Weizsäcker CF (1937) Phys Zeits 38:176; ibid. 39:633
Zee A (1973) Phys Rev D 7:3630
Zweig G (1964) An SU (3) model for strong interaction symmetry and its breaking. Preprints CERN-TH-401, 412. Reprinted in Developments in the quark theory of hadrons, vol 1. In Lichtenberg DB, Rosen SP (eds) Hadronic Press (1980), Nonantum, Massachusetts

Standard Model of Fundamental Interactions

9

Particle physicists have often come out with very imaginative linguistic creations. Asymptotic freedom and infrared slavery are good examples but also leptons and hadrons for matter particles. Gell-Mann was inspired by Joyce's novel *"Finnegans Wake"* for the naming of quarks and he is also credited for the name quantum chromodynamics. On the other hand, particle physicists have completely failed in finding a proper name for the theory of fundamental interactions that in principle describes all nongravitational physical phenomena from 10^{-19} m (resolution of the LHC) to at least 10^{11} m (distance earth-sun). The theory consisting of the electroweak gauge theory and the gauge theory of the strong interactions carries the fanciless name "Standard Model". Nobody in his right mind would nowadays talk about the Maxwell model of electrodynamics or Einstein's relativity model. But the comprehensive theory that agrees with all experimental findings over at least 30 (!) orders of magnitude in distance[1] is known by the mundane name Standard Model. The name goes back to the early 1970s when several competing models for the unification of the weak and electromagnetic interactions were on the market. As experiments more and more favoured one of those models, which was originally called Weinberg model, then Salam-Weinberg model and finally Glashow-Salam-Weinberg model, the theory was for simplicity called Standard Model. Even after QCD, which is as little model-like as QED, had passed all experimental tests with flying colours, the unimaginative name Standard Model (now of all fundamental interactions except gravity) was kept. Some particle physicists, especially among the older generation, prefer "Standard Theory" to "Standard Model", but also this name is not really a flash of genius and

---

[1]In the first instance, this applies to the luminous matter in the universe. Astrophysical analyses suggest the existence of "dark matter", which could require an extension of the Standard Model (Chaps. 10, 11).

© Springer Nature Switzerland AG 2019
G. Ecker, *Particles, Fields, Quanta*, Undergraduate Lecture Notes in Physics,
https://doi.org/10.1007/978-3-030-14479-1_9

has not been generally accepted. Thus, for lack of a better alternative, we also remain grudgingly with the Standard Model of the fundamental interactions in this book.

On the Way to the Standard Model

As discussed in the previous two chapters, all interactions relevant in the microcosm are described by non-abelian gauge theories (Yang-Mills theories). The strong interactions are mediated by eight gluons corresponding to the gauge group $SU(3)$. For the electroweak interactions at least three gauge bosons must be taken into account: W^+, W^- for the weak interaction and the photon for the electromagnetic interaction. The theoretician will then ask himself which non-abelian gauge groups are compatible with these three gauge bosons. The answer is unique, only the group $SU(2)$ satisfies all conditions. Rather exceptionally, nature does not content itself with the mathematically simplest possibility in this case but it insists on an additional (neutral) gauge boson for the electroweak interactions, the Z boson. But also for four gauge bosons there is only one (non-abelian) gauge group. Thus, the electroweak interactions are described by a gauge theory with the gauge group $SU(2) \times U(1)$.

The scientific papers of the subsequent Nobel Laureates Sheldon Glashow, Abdus Salam and Steven Weinberg (Glashow 1961; Salam 1968; Weinberg 1967), who all used the gauge group $SU(2) \times U(1)$, received little attention originally. In addition to purely theoretical questions like spontaneous symmetry breaking (Chap. 7) or renormalisability of the theory (Chap. 6), also an experimental aspect was responsible for this initial disregard. All known phenomena of the weak interactions like β decay were compatible with the existence of charged gauge bosons $W^\pm$ (Chap. 7). On the other hand, there were no experimental indications for the exchange of (neutral) Z bosons. In the physics jargon this circumstance was known as the absence of neutral weak currents, in contrast to the charged weak currents already known from Fermi theory that are induced by the exchange of charged W bosons according to modern view. The situation changed once neutrino beams became available at the beginning of the 1970s. In 1973 at CERN, both the inelastic neutrino-nucleon scattering $\nu_\mu + N \rightarrow \nu_\mu + X$ (Hasert et al. 1973b) (X stands for an arbitrary hadronic final state with the same charge as the initial nucleon N) and the elastic neutrino-electron scattering $\nu_\mu + e^- \rightarrow \nu_\mu + e^-$ (Hasert et al. 1973a) were detected. In the Standard Model, the latter purely leptonic process is represented by the Z-exchange diagram in Fig. 9.1 to lowest order in perturbation theory. With the discovery of neutral weak currents the acceptance of an electroweak gauge theory increased tremendously and the Glashow-Salam-Weinberg model became the Standard Model.

Overall, the Standard Model of the fundamental interactions is a Yang-Mills theory with the gauge group $SU(3) \times SU(2)_L \times U(1)$ where $SU(3)$ stands for QCD and $SU(2)_L \times U(1)$ for the electroweak interactions. The index L reminds us that the weak interactions distinguish between left and right (parity violation, Chap. 7). Group theoretically, the group $SU(2)_L$ is identical with $SU(2)$, which describes spin and

Fig. 9.1 Z-exchange diagram for elastic neutrino-electron scattering in Born approximation

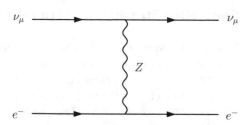

also Heisenberg's isospin. For that reason, $SU(2)_L$ is sometimes called the group of weak isospin.

Until the beginning of the 1970s, all known hadrons (particles and resonances subject to the strong interaction) could be explained as bound states of only three quarks (u, d, s; Chap. 8). On the other hand, there were already four leptons: e, ν_e, μ, ν_μ. For esthetic reasons, James Bjorken and Sheldon Glashow already suggested in 1964 that there should also be a fourth quark (Bjorken and Glashow 1964). A more substantial argument was put forward by Sheldon Glashow, Jean Iliopoulos and Luciano Maiani (Glashow et al. 1970). With the three known quarks, some experimentally observed selection rules for weak decays of kaons could not be explained. For the purpose of illustration, we consider two simple leptonic kaon decays. The charged kaon K^+ decays predominantly – with more than 60% probability – into a muon and the associated neutrino: $K^+ \rightarrow \mu^+ \nu_\mu$. On the other hand, the similar decay of the neutral kaon, $K_L^0 \rightarrow \mu^+ \mu^-$, is severely suppressed. A comparison between the two partial decay rates yields

$$\frac{\Gamma(K_L^0 \rightarrow \mu^+ \mu^-)}{\Gamma(K^+ \rightarrow \mu^+ \nu_\mu)} \simeq 3 \cdot 10^{-9} \,. \tag{9.1}$$

The suppression of the K_L^0 decay is an especially marked example for the general observation that neutral weak currents with a change of strangeness are strongly suppressed. Actually, in kaon decays the strangeness changes (almost) always. For our two leptonic decays this also is the case because kaons are "strange" mesons, which decay here into purely leptonic final states ($\mu^+ \mu^-$ and $\mu^+ \nu_\mu$, resp.). But leptons have no strangeness – in the quark model the quantum number strangeness is associated with the s quark. The K^+ decay is a normal weak process induced by charged weak currents, but the K_L^0 decay cannot occur at lowest order of perturbation theory because there are no neutral weak currents with change of strangeness (more generally, with change of quark flavour) in the Standard Model. Glashow, Iliopoulos and Maiani recognized that for such a selection rule (GIM mechanism) of weak decays a fourth quark is needed, now known as charm quark c. The discovery of a bound state $c\bar{c}$ of a charm quark and its antiparticle (Aubert et al. 1974; Augustin et al. 1974; Nobel Prize 1976 for Burton Richter and Samuel Ting) gave the Standard Model an additional boost. The GIM mechanism suppresses only neutral weak currents with a change of flavour because otherwise neutral weak currents (with conservation of flavour) could not have been detected at CERN in 1973.

Generation Structure of Matter Particles

With the discovery of the charm quark a generation structure of the fundamental fermions began to emerge. The first generation comprises those fermions that are of direct relevance for our environment: electrons and their neutrinos and the lightest quarks u, d making up the nucleons and thereby the atomic nuclei. That explains the reaction of the Nobel Laureate Isaac Rabi to the discovery of muons: "Who ordered that?" In a way, this question concerns the whole second generation: muons and their neutrinos and the quarks c and s.

For the theoretician it is especially important that both generations are complete in the following sense. So far, we have tacitly assumed that the symmetries of the field equations that can be read off most easily from the corresponding Lagrangian are not affected by quantum effects. So-called "anomalous" symmetries, which are modified by quantum effects, are relatively rare and they can actually have experimental consequences for global symmetries. But for gauge symmetries such anomalies would be fatal. It turns out that a consistent quantisation and renormalisation is only possible for anomaly-free gauge theories. In the framework of the Standard Model a gauge anomaly in principle only can occur where the interaction distinguishes between left and right, i.e. in the electroweak sector with both vector and axial-vector couplings. QCD is therefore not affected, it is a vectorial gauge theory and thus anomaly-free. For the electroweak gauge theory this is not the case automatically. The conditions for the absence of anomalies in this case are conditions for the quantum numbers of the fermions in the theory. In the case of the gauge group $SU(2)_L \times U(1)$ those conditions are reduced to a single one (Bouchiat et al. 1972; Gross and Jackiw 1972): the sum of all fermion charges must vanish where only those (left-handed) fermions are counted[2] that are in doublets of $SU(2)_L$. Let us then consider the particles in the first generation. The sum of the doublet charges in the quark sector gives $Q = \frac{2}{3}e - \frac{1}{3}e = \frac{1}{3}e$ because the u quark has charge $\frac{2}{3}e$ and the d quark $-\frac{1}{3}e$. With quarks alone we would therefore be in trouble. From the leptons in the first generation only the electron is relevant for counting because the neutrinos are electrically neutral. Thus the lepton sector contributes a charge $Q = -e$ and the sum of the lepton and quark charges still does not vanish. But we have forgotten to take into account that every quark comes in three colours, which have of course all the same electric charge. Hence the final counting is

$$Q_{\text{total}} = Q_{\text{quarks}} + Q_{\text{leptons}} = 3 \cdot \frac{e}{3} - e = 0 \,, \tag{9.2}$$

and the Standard Model with complete fermion generations is indeed anomaly-free.

Before we start pondering too much where nature has its knowledge about the quantisation of gauge theories from, let us reiterate the bare facts. That each generation contains the same number of quarks (namely two) already follows from the

---

[2]CPT invariance of the theory requires that with each particle also its antiparticle is represented in the theory. Therefore, the sum of all fermion charges vanishes in any quantum field theory.

GIM mechanism, which however does not lead to any relation between the quark and lepton sectors. Such a relation is implied by the condition for the absence of gauge anomalies, which in addition requires the existence of three colours. Why the number of quark colours is correlated with the electrical charges of the quarks, is another aspect that can make many a simple particle physicist dizzy. For the time being, we leave these questions to deeper thinkers (Chap. 10). In the meantime, we rejoice that the fermions appearing in nature allow for quantisation and renormalisation of our beautiful Standard Model.

The first two generations had just been completed when the next surprise followed. In 1975, another charged lepton was discovered at the e^+e^- collider SPEAR at the S(tanford)L(inear)A(ccelerator)C(enter) (Perl et al. 1975). Martin Perl, the leader of the experiment, received the Nobel Prize of 1995 for the discovery of the τ lepton. This particle was beyond any doubt a further so-called sequential lepton, i.e., except for its mass it was very similar to its siblings e and μ. With a mass of $m_\tau \simeq 1777\,\mathrm{MeV/c^2}$ it is however nearly twice as massive as a nucleon so at least for the linguist the name "lepton" appears to be a misnomer. A little later it turned out that there is also an associated neutrino ν_τ confirming the status of leptons. There was only one serious problem with those leptons. Where were the quarks of the third fermion generation that would reestablish the anomaly freedom of the Standard Model?

In 1973, the Japanese particle physicists Makoto Kobayashi and Toshihide Maskawa (Nobel Prize 2008) published an article (Kobayashi and Maskawa 1973) where they investigated several scenarios how the experimentally established CP violation (Chap. 4) could be implemented in an electroweak gauge theory. In the simplest version of spontaneously broken electroweak gauge symmetry (Chap. 7 and later in this chapter), as it is realised in the Standard Model, at least six quarks are necessary to implement CP violation. Together with the discovery of the τ lepton, this was a strong hint that the quarks of the third generation were still awaiting detection. And indeed, soon afterwards the bottom quark b with an electric charge $-e/3$ was found at the F(ermi)N(ational)A(ccelerator)L(aboratory) (Herb et al. 1977). At this point, a large majority of particle physicists was convinced that before long also a fermionic partner with charge $2e/3$ would be found. Actually it took almost 20 more years before the top quark t was discovered with the "correct" charge in two experiments at FNAL (Abachi et al. 1995; Abe et al. 1995).

There are good reasons to assume that there is no further generation of relatively "light" fermions. We will come back to this point shortly. We collect the three known generations of fundamental fermions in the two Tables 9.1 and 9.2. In the Standard Model, the masses of the fundamental fermions are not calculable but must be determined experimentally. At this time, the theory has no explanation to offer for the wide spectrum of quark and lepton masses. Leaving aside the neutrino masses (Chap. 10), the spectrum of fundamental fermions covers the range from 0.5 MeV (electron) to 170 GeV (top quark), corresponding to a difference of more than five orders of magnitude. All these masses are generated by the B(rout)E(nglert)H(iggs) mechanism (Chap. 7), but the specific values of the masses are free parameters in the Standard Model.

Table 9.1 Three generations of leptons (without antileptons). The bounds for neutrino masses are from particle decays. Much stronger bounds from the cosmic background radiation and from neutrino oscillations (Chap. 10) can be found in the Review of Particle Properties (Tanabashi et al. 2018)

Particle	Spin ($\hbar$)	Charge (Q/e)	Mass $\cdot\, c^2$ (MeV)
e^-	1/2	-1	0.5109989461(31)
ν_e	1/2	0	$<2 \cdot 10^{-6}$
μ^-	1/2	-1	105.6583745(24)
ν_μ	1/2	0	<0.19
τ^-	1/2	-1	1776.86(12)
ν_τ	1/2	0	<18.2

Of the gauge bosons, the photon and the gluons are massless, only the W and Z bosons have nonvanishing masses. One could be tempted to expect that in analogy to the fermion masses also M_W and M_Z are free parameters of the theory. But in fact, the Standard Model makes very precise predictions for these masses. In 1983, those predictions were beautifully confirmed by two experiments at CERN (UA1, Arnison et al. 1983; UA2, Banner et al. 1983; UA2, Bagnaia et al. 1983). The present values are (Tanabashi et al. 2018)

$$M_W = (80.379 \pm 0.012)\,\text{GeV/c}^2$$
$$M_Z = (91.1876 \pm 0.0021)\,\text{GeV/c}^2 \,. \tag{9.3}$$

The experimental confirmation of the theoretical predictions is one of the greatest successes of modern particle physics (Nobel Prize 1984 for Carlo Rubbia and Simon van der Meer). Why was it possible for the weak gauge bosons what is not possible for quarks and leptons? The keyword is gauge invariance. Both in QED and in QCD gauge invariance requires massless gauge bosons (photon, gluons). Although the electroweak gauge invariance is broken by the BEH mechanism, this symmetry breaking is a "soft" breaking in the terminology of quantum field theory. As explained in Chap. 7, the spontaneous symmetry breaking affects the ground state of the theory (vacuum), but the parameters of the theory are still constrained by gauge invariance as if the symmetry had not been broken at all. This is the reason why M_W and M_Z are calculable not only in the Born approximation but to all orders in perturbation theory. To date, these masses have been calculated up to and including the two-loop level. It cannot be emphasised too often, the theoretical values agree impressively well with the experimental results (9.3).

The gauge bosons W and Z are not only very massive but also highly unstable. With a simple argument (Bertlmann and Pietschmann 1977), the number of light neutrinos can be determined from a comparison between the experimentally measured and the theoretically calculated decay width of the Z boson. In the 1990s, the total decay width of the Z boson was measured very precisely in experiments at the L(arge)E(lectron)P(ositron) collider at CERN, the predecessor of the LHC. If

Table 9.2 Three generations of quarks (without antiquarks). Since quarks cannot be isolated as free particles (confinement), the listed masses depend on certain assumptions. Details can be found in the Review of Particle Properties (Tanabashi et al. 2018)

Particle	Spin ($\hbar$)	Charge (Q/e)	Mass $\cdot c^2$ (MeV)
u	1/2	2/3	$\simeq 2.2$
d	1/2	$-1/3$	$\simeq 4.7$
c	1/2	2/3	$1275(^{+25}_{-35})$
s	1/2	$-1/3$	$95(^{+9}_{-3})$
t	1/2	2/3	$173.0(0.4) \cdot 10^3$
b	1/2	$-1/3$	$4.18(4) \cdot 10^3$

the partial decay widths for the different decay channels of the Z boson are added, the sum differs from the measured total width. Within the Standard Model, this difference, also known as the "invisible" decay width, is due to the decays of the Z boson into neutrinos and their antiparticles, $Z \rightarrow \nu\bar{\nu}$. These decays do not leave any visible traces in the detectors. Due to the conservation of energy, such a decay can only occur if the mass of the specific neutrino (equal to the mass of the associated antineutrino) is smaller than $M_Z/2$. The comparison between theory and experiment then shows that there are exactly three "light" neutrinos (i.e. with $m_\nu < M_Z/2$), the known ν_e, ν_μ and ν_τ. Making the natural assumption that each neutrino is accompanied by a charged lepton as partner, the anomaly freedom of the Standard Model implies that there are only the three fermion generations already known (Tables 9.1, 9.2). Incidentally, this experimental evidence for three light neutrinos also agrees with indications from astrophysics and cosmology.

Higgs Sector of the Standard Model

Towards the end of the last century, the structure and the ingredients of the Standard Model were by and large known. However, an experimental confirmation of the mechanism of spontaneous breaking of the electroweak gauge symmetry was still missing. This mechanism must provide the three degrees of freedom that make three massive vector bosons W^+, W^-, Z out of massless gauge bosons (Chap. 7). In the Standard Model the simplest version of the BEH mechanism seems to be realised. By an appropriately constructed Higgs potential, a complex $SU(2)_L$ doublet of scalar fields is forced to furnish in the ground state a constant value $v \simeq 246\,\mathrm{GeV}$ (also known as Fermi scale) for the vacuum expectation value of the electrically neutral partner in this doublet. All masses in the Standard Model, for W and Z as well as for leptons and quarks, are proportional to the Fermi scale.

A complex doublet of scalar fields has four real degrees of freedom. Three of them are "eaten" up by the gauge bosons $W^\pm$, Z to become massive. Thus, there is one real scalar field left and the associated particle is the (in)famous Higgs boson.[3] In addition to the Fermi scale, there is one more free parameter in the Higgs potential and therefore the mass of the Higgs boson is not calculable in the Standard Model just like the fermion masses. On the other hand, the couplings of the Higgs boson to all other particles in the Standard Model, gauge bosons as well as fermions, are fixed by the masses of those particles. In particular, it turns out that the Higgs boson prefers to decay into the heaviest particles as long as the conservation of energy does not prevent the decay. The hope of particle physicists that the Higgs boson could be detected in the LEP experiments turned out to be premature. After the shutdown of LEP in 2001, an allowed domain (with 95% probability)

$$114\,\text{GeV/c}^2 < M_H < 144\,\text{GeV/c}^2 \tag{9.4}$$

remained for the Higgs mass. The lower bound is a simple consequence of the fact that LEP did not provide enough energy to produce a Higgs boson, but where does the upper bound come from? As all other particle masses, the Higgs mass enters in the calculation of many observable quantities. From the comparison of theoretical predictions for those observables with results from the LEP experiments an allowed domain for M_H can be obtained, in particular the upper bound $144\,\text{GeV/c}^2$. Incidentally, the allowed domain (9.4) implies that the two-body decay channels with the heaviest particles ($H \rightarrow t\bar{t}, W^+W^-, ZZ$) are all excluded because due to energy conservation the sum of the masses in the final state must always be smaller than M_H.

The experimental clarification of the spontaneous breaking of electroweak gauge symmetry was the main motivation for the construction of the LHC at CERN. After a delay of one year due to a technical defect, the LHC went into operation in November 2009. Following a long-time tradition at CERN, accelerator and detectors have been working without problems ever since and actually better than expected. In the first phase until the beginning of 2013, protons were colliding with a center-of-mass energy of seven and then eight TeV. The first results were summarised by the participating experimentalists as follows: "We have rediscovered the Standard Model at the LHC."

In contrast to LEP, the available energy at the LHC is big enough to produce a Higgs boson with a mass in the domain (9.4). The key question was therefore whether this needle in a haystack (compare, e.g., Fig. 4.1) could actually be found. In a joint meeting in the CERN Auditorium on July 4, 2012, the two big LHC experiments ATLAS and CMS presented their results for the existence of a boson in the expected

---

[3]In the popular literature the Higgs boson is sometimes referred to as the "God particle". This goes back to a book of Lederman and Teresi with the title "The God Particle: if the Universe is the Answer, What is the Question?". According to well-founded rumours, Leon Lederman had originally proposed the title "The Goddamn Particle" to express already in the title his frustration that the particle stubbornly eluded experimental detection. Unfortunately, the publisher insisted for obvious reasons on the actual, less appropriate title.

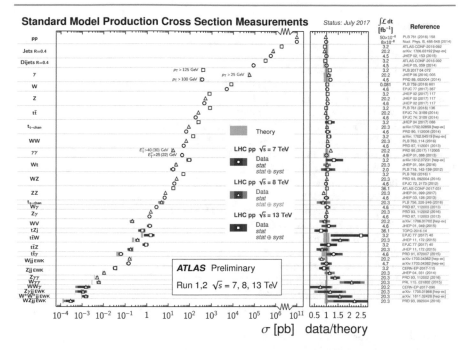

Fig. 9.2 Comparison between theory and experiment for several cross sections measured in the ATLAS experiment at LHC; status of July 2017 with proton-proton center-of-mass energies up to 13 TeV (From Tanabashi et al. 2018; with kind permission of © Particle Data Group 2018. All Rights Reserved)

region[4] (9.4). Since then it has been shown in painstakingly detailed analyses that this boson with a mass (Tanabashi et al. 2018)

$$M_H = (125.18 \pm 0.16)\,\text{GeV/c}^2 \qquad (9.5)$$

has indeed the expected properties of the Higgs boson, in particular the decay probabilities into various final states. The Royal Swedish Academy of Sciences was also quickly convinced and it bestowed the Nobel Prize for Physics of the year 2013 to François Englert and Peter Higgs.

Except for neutrino masses (Chap. 10), all parameters of the Standard Model are now known. How well the Standard Model works also at LHC energies, can be seen for instance from comparison between theory and experiment for some cross sections measured at the LHC in Fig. 9.2. Are we then again in a similar situation as before 1900 that "all the important basic laws and facts of particle physics have already been discovered"? The final two chapters will address this question.

---

[4]Unlike for most election forecasts, one can in general rely on statements of probability in particle physics.

References

Abachi S et al (1995) Phys Rev Lett 74:2422
Abe F et al (1995) Phys Rev Lett 74:2626
Arnison G et al (UA1 Collaboration) (1983) Phys Lett B 122:103; ibid. B 126:398
Aubert JJ et al (1974) Phys Rev Lett 33:1404
Augustin JE et al (1974) Phys Rev Lett 33:1406
Bagnaia P (UA2 Collaboration) et al (1983) Phys Lett B 129:130
Banner M (UA2 Collaboration) et al (1983) Phys Lett B 122:476
Bertlmann R, Pietschmann H (1977) Phys Rev D 15:683
Bjorken JD, Glashow SL (1964) Phys Lett 11:255
Bouchiat C, Iliopoulos J, Meyer P (1972) Phys Lett B 38:519
Glashow SL (1961) Nucl Phys 22:579
Glashow SL, Iliopoulos J, Maiani L (1970) Phys Rev D 2:1285
Gross DJ, Jackiw R (1972) Phys Rev D 6:477
Hasert FJ et al (1973a) Phys Lett B 46:121
Hasert FJ et al (1973b) Phys Lett B 46:138
Herb SW et al (1977) Phys Rev Lett 39:252
Kobayashi M, Maskawa T (1973) Prog Theor Phys 49:652
Perl ML et al (1975) Phys Rev Lett 35:1489
Salam A (1968) In: Svartholm N (ed) Proceedings of 8th Nobel symposium. Almqvist and Wiksell, Stockholm
Tanabashi M et al (Particle Data Group) (2018) Phys Rev D 98:030001
Weinberg S (1967) Phys Rev Lett 19:1264

Beyond the Standard Model?   10

The Standard Model of fundamental interactions describes all nongravitational physical phenomena. Nevertheless, there is a general consensus among particle physicists that the Standard Model is not the final theory of electromagnetic, weak and strong interactions.

Massive Neutrinos

A small but important extension of the original Standard Model concerns neutrino masses. Soon after the discovery of the electron neutrino it was clear that the mass of the neutrino must be significantly smaller than the electron mass. Therefore, the Standard Model as originally conceived had all neutrinos massless. Together with the minimal Higgs sector (Chap. 9), this feature was implemented in the Standard Model by allowing only two degrees of freedom for each neutrino (V–A structure: left-handed neutrinos, right-handed antineutrinos). In contrast, all charged matter particles have four degrees of freedom because they are massive (Dirac equation, Chap. 3).

Our atmosphere buzzes with neutrinos. In addition to cosmic background neutrinos, there are two main sources: neutrinos coming from the sun and neutrinos produced in the atmosphere through decays in cosmic rays. In spite of the enormous neutrino flux from the sun ($6.6 \cdot 10^{14}$ neutrinos/m^2 s), detection is difficult due to the tiny cross sections. In an experiment in the Homestake gold mine in South Dakota, Raymond Davis (Nobel Prize 2002) succeeded to detect solar neutrinos for the first time. But already the first results in 1968 only yielded about one third of the expected neutrino flux (Davis et al. 1968). After both the experiment and theoretical calculations of the solar neutrino flux (Standard Solar Model SSM, Bahcall

© Springer Nature Switzerland AG 2019   97
G. Ecker, *Particles, Fields, Quanta*, Undergraduate Lecture Notes in Physics,
https://doi.org/10.1007/978-3-030-14479-1_10

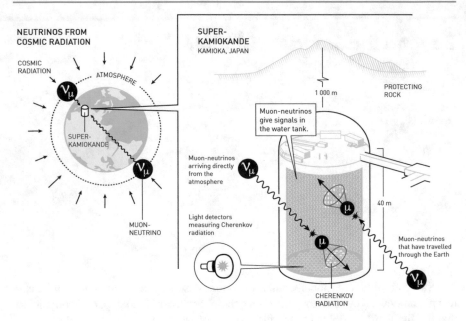

Fig. 10.1 Super-Kamiokande neutrino experiment for atmospheric neutrinos (From http://www. nobelprize.org; with kind permission from © Johan Jarnestad/The Royal Swedish Academy of Sciences. All Rights Reserved)

et al. 2005) had been substantially refined over a period of more than 25 years, it became clear that the measured flux was indeed only about 30% of the calculated flux. This was confirmed in 1989 by the Kamiokande experiment (Nobel Prize 2002 for Masatoshi Koshiba) by measuring the elastic neutrino electron scattering induced by solar neutrinos (Hirata et al. 1989).

It was obvious that something was happening to the electron neutrinos on their way from the sun to the earth. The most elegant and ultimately correct explanation is due to the quantum theoretical phenomenon of neutrino oscillations. On the way from the production to the detector, a neutrino of a certain type (flavour e, μ or τ) turns into a superposition of all three neutrino flavours. A necessary condition for such oscillations to occur, however, is that unlike in the original Standard Model the neutrinos must have different masses.

The ultimate confirmation of neutrino oscillations came from two experiments led by Takaaki Kajita and Arthur McDonald (Nobel Prize 2015). In the Super-Kamiokande experiment (Fukuda et al. 1998) in Kamioka, Japan, muon and electron neutrinos generated in the earth atmosphere were detected in a giant water tank, 1000 m below the earth's surface (Fig. 10.1). While for the electron neutrinos it made practically no difference whether they came directly from the atmosphere or whether they traversed the whole earth before entering the detector from below, there was a significant difference for muon neutrinos. The interpretation was straightforward.

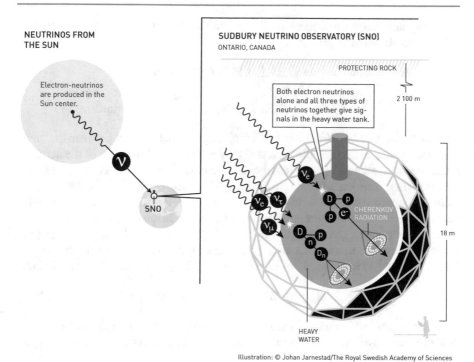

NEUTRINOS FROM
THE SUN

Electron-neutrinos
are produced in the
Sun center.

SNO

SUDBURY NEUTRINO OBSERVATORY (SNO)
ONTARIO, CANADA

PROTECTING ROCK

Both electron neutrinos
alone and all three types of
neutrinos together give signals in the heavy water tank.

2 100 m

CHERENKOV
RADIATION

18 m

HEAVY
WATER

Illustration: © Johan Jarnestad/The Royal Swedish Academy of Sciences

Fig. 10.2 Sudbury Neutrino Observatory for solar neutrinos (From http://www.nobelprize.org; with kind permission from © Johan Jarnestad/The Royal Swedish Academy of Sciences. All Rights Reserved)

Whereas the muon neutrinos from "above" had to travel only a few kilometers to the detector, those from "below" had a substantially longer itinerary. Thus, they had ample opportunities to oscillate especially into τ neutrinos, which could not be detected in the experiment.

The last piece of the solar neutrino puzzle came from the Sudbury Neutrino Observatory (Fig. 10.2), an experiment in Ontario, Canada with a tank of 1000 tons of heavy water, some 2100 m below the earth's surface. In that experiment, the complete neutrino flux from the sun could be measured, both the electron neutrinos and the muon and τ neutrinos generated by oscillations (Ahmad et al. 2002). As expected (Homestake experiment, Davis et al. 1968), only one third of the flux consists of electron neutrinos but the total measured neutrino flux agrees with the prediction of the SSM.

As mentioned above, neutrino oscillations can only occur if the neutrinos are nondegenerate, i.e., if their masses are different. Thus, the experimental verification of neutrino oscillations implies that the Standard Model in its original form with massless neutrinos must be modified. However, oscillation data only contain information about differences of (the squares of) neutrino masses. Denoting the neutrinos with definite masses as ν_1, ν_2, ν_3, a global analysis of all oscillation experiments, including also reactor and accelerator experiments, yields the following up-to-date values (Tanabashi et al. 2018):

$$\Delta_{21}^2 = |m_2^2 - m_1^2| = (7.53 \pm 0.18) \cdot 10^{-5} \text{ eV}^2/c^4$$

$$\Delta_{32}^2 = |m_3^2 - m_2^2| = \begin{cases} (2.51 \pm 0.05) \\ (2.56 \pm 0.04) \end{cases} \cdot 10^{-3} \text{ eV}^2/c^4 \begin{array}{l} \text{(normal)} \\ \text{(inverse)} \end{array}. \quad (10.1)$$

At present (September 2018), one cannot distinguish definitively between the "normal" ($m_3 > m_2 > m_1$) and the "inverse" ($m_2 > m_1 > m_3$) ordering of the neutrino mass spectrum although global fits of available experimental results favour the normal ordering. This explains the two slightly different values for Δ_{32}^2. In addition, from the experimental results the so-called leptonic mixing angles that characterise the superpositions of the flavour eigenstates ν_e, ν_μ, ν_τ in the mass eigenstates ν_1, ν_2, ν_3 can be extracted. From the results in Eq. (10.1) one deduces that at least two of the three neutrinos are massive. Denoting the mass of the heaviest neutrino as m_h and that of the lightest neutrino as m_l, the results of (10.1) give rise to the lower bounds

$$m_h \geq 0.051 \text{ eV}/c^2, \qquad m_l \geq 0. \quad (10.2)$$

Since neutrinos influence the formation of structures in the early universe, astroparticle physics provides also an upper bound for the sum of neutrino masses (Ade et al. 2016):

$$\sum_i m_i < 0.23 \text{ eV}/c^2. \quad (10.3)$$

Together, the results (10.1) and (10.3) give rise to the bounds[1]

$$0.051 \leq m_h c^2/\text{eV} < 0.088$$
$$0 \leq m_l c^2/\text{eV} < 0.071. \quad (10.4)$$

The great successes of neutrino physics during the last years have raised many new questions. On the experimental side, the absolute measurement of at least one neutrino mass is still missing. Experiments in preparation intend to measure the electron spectrum in tritium β decay.[2] The expected sensitivity of $\sim 0.2 \text{ eV}/c^2$ for the neutrino mass should at least approach the currently allowed domain (10.4). Another question still open is whether neutrinos are their own antiparticles (Majorana neutrinos) or not (Dirac neutrinos). On the theoretical side, there are many proposals for extending the Standard Model to understand the mass spectrum of neutrinos and the associated mixing angles. However, at present masses and mixing angles in the lepton sector are as unexplained as the analogous quantities in the hadron sector (quark masses and weak mixing angles of the Cabibbo-Kobayashi-Maskawa mixing matrix).

---

[1]Upper bounds are given for normal ordering only; they are a little tighter for inverse ordering.
[2]The KATRIN experiment in Karlsruhe started data taking in May 2018.

Extensions of the Standard Model

The Standard Model began with the successful unification of electromagnetic and weak interactions in the framework of a gauge theory and the subsequent realisation that also the strong interactions can be described by a gauge theory. The Standard Model is based on the gauge group $SU(3) \times SU(2)_L \times U(1)$. The product structure of this gauge group tells the particle physicist that the Standard Model has three independent gauge coupling constants (see also Chap. 8). For evident reasons, they are usually denoted as g_1, g_2, g_3. Here, g_1 and g_2 are combinations of the elementary charge e and of the weak coupling constant g_W (Fig. 7.3); $g_3 = g_s$ (Fig. 8.5) is the coupling constant of QCD. Against the background of a common gauge structure, the idea of a further unification of the three interactions was already envisaged in the first half of the 1970s. In such a scheme, the Standard Model would be the "low-energy version" of a more fundamental theory with a (in the technical sense) simple gauge group with a single coupling constant.

Before we pursue the idea of grand unification usually abbreviated as GUT for Grand Unified Theory, we discuss a few additional arguments suggesting an extension of the Standard Model.

- In addition to the three gauge coupling constants, the Standard Model possesses several other free parameters. Together with the gauge couplings, the original Standard Model with massless neutrinos has 18 free parameters,[3] which are in particular responsible for the fermion masses (quarks and leptons) and for the weak mixing angles (CKM matrix) mentioned above. All those parameters are not constrained by the theory and must be determined experimentally.
- With the neutrino masses the hierarchy of fermion masses becomes even more mysterious. Using the maximal neutrino mass in Eq. (10.4), the masses of matter particles encompass at least 12 orders of magnitude:

$$\frac{m_{\text{top}}}{m_\nu} > 2 \cdot 10^{12} . \tag{10.5}$$

Although it may seem to be paradoxical at first sight, grand unification at high energies furnishes a possible explanation for the tiny neutrino masses (see below).
- In the Standard Model, spontaneous breaking of the gauge symmetry (BEH mechanism) is not only responsible for masses and mixing angles, but it also provides a parametrisation of CP violation (Chap. 9). This mechanism of CP violation is compatible with all experimental results in the hadron sector.[4] However, the experts agree that the mechanism is not sufficient to explain the asymmetry between matter and antimatter in the universe.

---

[3]In the extended Standard Model with massive neutrinos, there are in total 25 free parameters with Dirac neutrinos, 27 with Majorana neutrinos.

[4]The possible CP violation in the lepton sector in extensions of the Standard Model with massive neutrinos is not yet confirmed definitively although some experimental indications have been found.

- Astrophysical findings, especially measurements of the cosmic microwave background (CMB) and the accelerated expansion of the universe (Nobel Prize 2011 for Saul Perlmutter, Brian Schmidt and Adam Riess), suggest that only about 5% of the energy density of the universe consists of normal matter (atoms). For the remaining 95% (about 26% "dark matter" that neither emits nor absorbs light, and 69% "dark energy") the Standard Model has no explanation. Most likely, gravity is responsible for dark energy, e.g., with a nonvanishing cosmological constant in Einstein's equations of general relativity.

Frank Wilczek formulated a persuasive motivation for searching for an extension of the theory (Wilczek 2016): "Since this theory is close to Nature's last word, we should take its remaining esthetic imperfections seriously." In fact, some of the arguments listed above may well belong to the category of "esthetic imperfections".

But in addition to esthetic imperfections, discrepancies between theoretical predictions and experimental results would be even more convincing arguments for an extension of the Standard Model. Considering the wealth of observables for which the Standard Model makes concrete predictions, it is statistically extremely unlikely that all those predictions agree exactly with experimental data. To assess discrepancies between theory and experiment, a careful analysis of the statistical significance is therefore mandatory.

To illustrate the problem, we consider a recent example. In December 2015, the two big LHC experiments ATLAS and CMS announced evidence for the existence of a new particle that decays into two photons. This particle, which for reasons soon to become clear has no official name so that we simply call it X particle here, would have had a mass of about $750\,\text{GeV}/c^2$, about six times as massive as the Higgs boson. This announcement hit the scene like a bomb. Until summer 2016, more than 500 theoretical articles appeared trying to explain which role such a particle could play in extensions of the Standard Model. Actually, in many of those papers as well as in conferences and workshops the insufficient statistical evidence of the data was emphasised. More concretely, this evidence is quantified by the experimentalists in the following way. In case of the X particle, the observable quantity is (a bit simplified) the number of measured photon pairs with an invariant mass (Chap. 1) around $750\,\text{GeV}/c^2$ minus the number of pairs predicted by the Standard Model in the relevant mass region. The experimental result is then announced in the form $\mu \pm \delta$ where μ is the statistical mean value and δ the experimental uncertainty. Now the crucial question is how significantly the result deviates from the Standard Model prediction that this difference between measured and predicted values should be exactly zero. In a compact way the answer is that the experimental result deviates with $\delta/|\mu|$ standard deviations ($\frac{\delta}{|\mu|}\sigma$) from the null hypothesis of the Standard Model. If we further assume that the data satisfy at least approximately a Gaussian distribution, probability theory furnishes the probability that the actual result lies between $\mu - n\,\delta$ and $\mu + n\,\delta$. For instance, for $n = 1$ (one standard deviation) this probability is 68%, while a probability of 99.9% corresponds to about $n = 3.3$ ($3.3\,\sigma$).

At the Spring Conferences of 2016, the experimenters announced a (so-called local) significance of $3.9\,\sigma$ (ATLAS) and $3.4\,\sigma$ (CMS), respectively. Remarkably

Fig. 10.3 Contribution of
hadronic vacuum
polarisation of lowest order
to the anomalous magnetic
moment of the muon

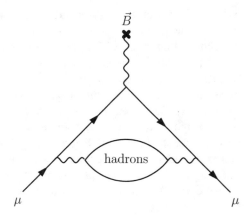

enough, the two experiments had independently found practically the same value for the mass of the presumptive X particle. At the High Energy Physics Conference in Chicago the latest results were announced on Aug. 5, 2016. Those results were based on substantially more data than before and they were completely unambiguous: the X particle had disappeared in the statistical noise and was officially declared dead. Was this a big blunder for experimental particle physics? Not at all! In the experimental presentations it had always been emphasised that according to a tacit agreement among particle physicists an official discovery requires a statistical significance[5] of at least $5\,\sigma$. Of course, that agreement had also been respected before the official announcement of the discovery of the Higgs boson in July 2012. Also the "explanations" of the X particle presented in theoretical articles should not be seen as a lack of credibility, but they document on the contrary a richness of imagination and the enthusiasm of particle physicists. There remains a certain bewilderment about the apparent caprioles of statistics that the deceptive fluctuations had occurred in both experiments at practically the same mass.

Although less spectacular than the deceased X particle, there are several other cases where a discrepancy between theory and experiment has been observed. However, in all those cases the statistical significance is smaller than $4\,\sigma$. To this date (September 2018), there is no compelling evidence for "New Physics" beyond the Standard Model. An interesting example with significance between $3.5\,\sigma$ and $4\,\sigma$ is the anomalous magnetic moment of the muon that was last measured in a precision experiment at the Brookhaven National Laboratory (Chap. 2). Different from the electron case (Chap. 5), with the present experimental accuracy not only electromagnetic corrections, but also contributions from the weak and strong interactions have to be taken into account for the anomalous magnetic moment of the muon. Whereas the weak contributions are sufficiently well known (including two-loop corrections), this is not the case for the strong interactions because not all corrections can be

---

[5]If this condition of significance would hold for opinion polls, most institutes in the business would have to close down.

captured in perturbation theory. In particular, this is true for the contribution of hadronic vacuum polarisation of lowest order depicted in Fig. 10.3. This hadronic correction is dominated by contributions at low energies where QCD perturbation theory is not applicable because of confinement. However, this contribution can be related to the cross section $\sigma(e^+e^- \to$ hadrons) and calculated with the help of experimental data. Even more difficult to calculate, albeit a bit less significant numerically at the present level of experimental accuracy, is the so-called hadronic light-by-light scattering. Therefore, intensive attempts are under way to improve the accuracy substantially, both theoretically (dispersion relations, lattice gauge theory) and experimentally. In particular, two new experiments (at Fermilab and at J-PARC in Japan) hope to improve the experimental accuracy by a factor of four. First results are to be expected soon, especially from the Fermilab experiment, which has already started operating.

Grand Unification

Because at this time there are no compelling reasons for a major extension of the Standard Model we return to its "esthetic imperfections". Since the discovery of asymptotic freedom in QCD (Chap. 8) the actual values of coupling constants are known to depend on the energy at which they are measured (see Fig. 8.5). For sufficiently small coupling constants this energy dependence can be calculated in perturbation theory. All those particles (fields) that are affected by the interaction(s) in question and whose mass ($\times c^2$) is smaller than the energy under consideration must be considered in such a calculation. If the experimental values of the coupling constants at some energy are known, e.g., at $E \sim 100\,\text{GeV}$, their energy dependence is calculable for all higher energies as long as no new particles with larger masses enter the game – and, of course, as long as no new interactions intervene. In Fig. 10.4 the energy dependence of the three gauge coupling constants g_1, g_2, g_3 is shown. More precisely, the inverse quantities α_i^{-1} are displayed where $\alpha_i = g_i^2/(4\pi\hbar c)$ ($i = 1, 2, 3$) are generalised fine-structure constants. The gauge couplings g_2 and g_3 for the non-abelian groups $SU(2)$ and $SU(3)$ decrease with increasing energy whereas the coupling g_1 of the abelian group $U(1)$ increases.

Looking first at the left plot in Fig. 10.4, we observe that the three coupling constants approach each other at energies 10^{14} to $10^{16}\,\text{GeV}$ even if there does not seem to be a common point of intersection. The first reaction of an unbiased observer is that there may well be a realistic basis for the presumed grand unification of the three fundamental interactions. This first impression is corroborated by recalling that the calculation for the left plot assumes that nothing new happens between $100\,\text{GeV}$ and $10^{16}\,\text{GeV}$, i.e., that there are no new interactions and no heavy matter particles in this enormous energy range. This scenario of the "great desert" has absolutely no theoretical basis, it just expresses our ignorance about physics at smaller distances. Thus, there are various approaches how one could cause the three gauge couplings to meet at the same point.

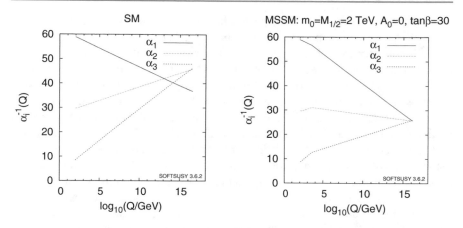

Fig. 10.4 Energy dependence of the quantities α_i^{-1} where $\alpha_i = g_i^2/(4\pi\hbar c)$ ($i = 1, 2, 3$) are generalised fine-structure constants. The three coupling constants stand for the three factors of the Standard Model gauge group $SU(3) \times SU(2)_L \times U(1)$. The decadic logarithms of the energy Q/GeV are displayed on the x-axis. For instance, the value 15 stands for an energy 10^{15} GeV. Left plot: Standard Model with "great desert". Right plot: MSSM with SUSY particles with masses of the order TeV/c^2 (From Tanabashi et al. 2018; with kind permission of © Particle Data Group 2018. All Rights Reserved)

The by far most popular attempt to make the desert bloom carries the name MSSM (Minimal Supersymmetric Standard Model). Supersymmetry is an unusual symmetry that relates bosons and fermions. In particular, supersymmetry (SUSY) requires that bosons and fermions are degenerate, i.e., for each boson there is a fermion with the same mass and vice versa. Such degenerate pairs do not exist in the known particle spectrum and thus SUSY would have to be a strongly broken symmetry. Since the first construction of a SUSY quantum field theory in the first half of the 1970s, the search for SUSY partners of the known gauge bosons and matter particles has started anew with each new accelerator. Until today (September 2018), also the LHC has not found any SUSY particles with masses smaller than about one TeV/c^2. Slowly but steadily, this puts the MSSM in difficulty as an explanation of grand unification because the SUSY particles must not be much more massive than one TeV/c^2 to make the three gauge couplings meet at about 10^{16} GeV (right plot in Fig. 10.4). The true SUSY aficionados are not really discouraged by this argument, nor by the petty objection that the Standard Model has about 20 free parameters, while the MSSM has more than 100. After all, it took 48 years from the BEH mechanism to the discovery of the Higgs boson whereas "only" 44 years have passed (Ellis 2015) since the first SUSY quantum field theory (Wess and Zumino 1974).

But quite independently of how the desert can be made to bloom, grand unification remains a fascinating idea. The most attractive version of a GUT is a gauge theory based on the gauge group $SO(10)$ that contains the group $SU(3) \times SU(2)_L \times U(1)$ of the Standard Model as a subgroup. In the Standard Model the matter particles of one generation are organised in different multiplets: $SU(3)$ triplets and singlets and altogether five (or even six including right-handed neutrinos to account for neutrino masses) independent doublets and singlets of $SU(2)_L$. The GUT group $SO(10)$ puts all those fermions of one generation into a single (irreducible) 16-dimensional representation. This implies in particular that, while it is a possibility in the Standard Model, a right-handed neutrino field is a necessity in an $SO(10)$ GUT allowing for massive neutrinos. But grand unification with a simple group like $SO(10)$ has still other attractive features. Whereas in the Standard Model the electric charges of particles are largely arbitrary,[6] the relative charges in a GUT are fixed by the group structure. In other words, if we define for instance the charge of the electron to be -1 (in units of e) in the usual convention, the charges of all other matter particles are fixed including the peculiar quark charges. But in addition the 16-dimensional representation for the matter particles also requires exactly three colours for each quark flavour. Finally, the group $SO(10)$ even answers the question raised in Chap. 9 how nature knows that a gauge theory must not have any (gauge) anomalies for a consistent quantisation and renormalisation. The group $SO(10)$ belongs to the class of anomaly-free groups and with the particle content of the 16-dimensional fermion representation also the Standard Model is guaranteed to be free of anomalies.

Grand unification and an $SO(10)$ GUT in particular would certainly entail a deeper understanding of the structure of the microcosm. But can it also be experimentally verified? After all, GUTs are expected to become relevant only for energies of the order of 10^{16} GeV that we will certainly not be able to reach in the foreseeable future. In addition to the 12 known gauge bosons (eight gluons, W^+, W^-, Z, γ), each GUT contains additional gauge fields most of which can induce baryon number-violating interactions because quarks and leptons of a fermion generation are contained in the same multiplet. In contrast, baryon number is absolutely conserved in the Standard Model because the 12 gauge fields of the Standard Model act separately on leptons and quarks. Therefore, the proton is absolutely stable in the Standard Model because there are no lighter baryons into which it could decay. In GUTs in general and in the $SO(10)$ gauge theory in particular, the situation is different. Since experiments have never detected a proton decay despite intensive searches, such baryon number-violating decays must be heavily suppressed. Just as the weak interactions are weak at low energies because the W and Z bosons are relatively massive, the gauge bosons of an $SO(10)$ GUT responsible for baryon number violation must be very massive in order to suppress the possible decays of the proton. For example, the experimental findings (Tanabashi et al. 2018) that the (partial) lifetime for the decay $p \rightarrow e^+ \pi^0$ is longer than $1.6 \cdot 10^{34}$ years[7] imply that the masses of the corresponding gauge bosons ($\times c^2$) cannot be much smaller than the unification scale 10^{16} GeV. If the dedicated experiments looking for proton decay would be able to increase their sensitivity still a little, the idea of grand unification could soon be testable experimentally.

---

[6]The abelian group $U(1)$ is the culprit.

[7]In quantum theory this statement is meaningful although only about $14 \cdot 10^9$ years have passed since the Big Bang. In case of doubt, consult a quantum physicist of your choice.

Einstein's dream to unify gravity with electromagnetism has been superseded in modern particle physics by the grand unification of the three fundamental interactions of the microcosm. However, the GUT scale of about 10^{16} GeV is not too far away from the Planck scale $E_P \sim 10^{19}$ GeV where quantum gravity should come into play (Appendix A). Of course, this relative proximity of the two scales keeps stimulating the phantasy of physicists.

Grand unification cannot answer all questions particle physicists like to pose. Except for the small neutrino masses, which find a natural explanation in $SO(10)$ gauge theories (seesaw mechanism, Minkowski 1977), GUTs do not offer any convincing explanation for the hierarchy of fermion masses. Also the question why there are exactly three fermion generations remains unanswered for the time being. But maybe superstring theory offers an admittedly unconventional answer to those questions. For many years it was the declared aim of string theorists to find a unique answer in their search for the Theory of Everything. Although even today there is no general consensus what superstring theory really is, the hypothesis is gaining ground that instead of a single one there could be an enormous multitude of different superstring theories (the inconceivable number 10^{500} is sometimes put forward), most of which or maybe all of which could be realised in causally disconnected regions of the universe. In all versions of that theory the fundamental parameters like masses and coupling constants would be different. Our part of the universe is only distinguished in that our set of fundamental parameters allows for the existence of life and therefore for the existence of physicists[8] who can pose such questions. But since there may be 10^{500} other possibilities it would be completely pointless to try to explain those parameters. Even among superstring disciples there are different views how far physics is sliding into the realm of metaphysics here.

References

Ade PAR et al (2016) Planck Collaboration. Astron Astrophys 594:A13. https://arxiv.org/abs/1502.01589

Ahmad QR et al (2002) SNO Collaboration. Phys Rev Lett 89:011301. http://arxiv.org/abs/nucl-ex/0204008

Bahcall JN, Serenelli AM, Basu S (2005) Astrophys J 621:L85. http://arxiv.org/abs/astro-ph/0412440

Davis R Jr, Harmer DS, Hoffman KC (1968) Phys Rev Lett 20:1205

Ellis J (2015) J Phys Conf Ser 631:012001. http://arxiv.org/abs/1501.05418

Fukuda Y et al (1998) Super-Kamiokande Collaboration. Phys Rev Lett 81:1562. http://arxiv.org/abs/hep-ex/9807003

Hirata KS et al (1989) Phys Rev Lett 63:16

Minkowski P (1977) Phys Lett B 67:421

Tanabashi M et al (2018) Particle Data Group. Phys Rev D 98:030001

Wess J, Zumino B (1974) Phys Lett B 49:52

Wilczek F (2016) Phil Trans Royal Soc A 374:20150257. http://arxiv.org/abs/1512.02094

---

[8]This so-called anthropic principle has a much longer history than superstring theory.

Outlook

11

In spite of the impressive confirmation of the Standard Model by LHC experiments, experimental and theoretical attempts will continue to look for hints for "New Physics" beyond the Standard Model. On the experimental side, these attempts can be characterised by three frontiers. In Fig. 11.1 those frontiers are shown graphically: high energies, high intensities and cosmic frontier. As also shown in the figure, the three frontiers overlap to some extent. The big LHC experiments ATLAS and CMS are mainly concerned with physics at the high-energy frontier. But the LHC does not only provide very high energies but also very high intensities. Thus, the smaller LHC experiment LHCb investigates decays of B particles (particles with at least one b quark) and even decays of the relatively light K mesons. On the other hand, both at the LHC and in specialised experiments of astroparticle physics searches for dark matter will be pursued.

Energy Frontier

This is not the place for a comprehensive overview of the present status of experimental high energy physics.[1] Instead we are going to highlight a few prominent developments that can be expected to shape particle physics in the next ten or more years. The primacy of highest energies will remain in Europe for some time to come. As already decided by the CERN Council, the luminosity of the LHC, a measure for the intensity of the two proton beams, will be raised in two steps by a factor between five and seven till 2026, in particular by installing new superconducting magnets with

---

[1]An up-to-date overview can be found in the proceedings of the biannual High Energy Physics Conferences, e.g., from the most recent conference in Seoul (http://www.ichep2018.org/).

© Springer Nature Switzerland AG 2019

G. Ecker, *Particles, Fields, Quanta*, Undergraduate Lecture Notes in Physics,
https://doi.org/10.1007/978-3-030-14479-1_11

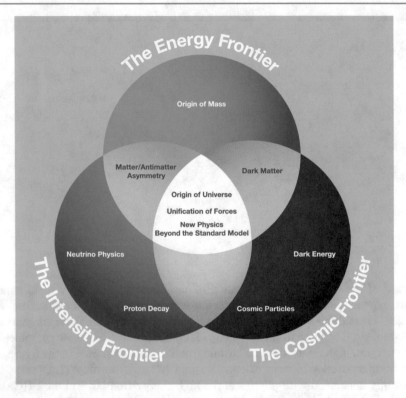

Fig. 11.1   Three frontiers of present-day experimental high energy physics: energy frontier, intensity frontier and cosmic frontier. (From Particle Physics Project Prioritization Panel (P5), U.S. Particle Physics: Scientific Opportunities; with kind permission of © U.S. Department of Energy, Office of Science 2008. All Rights Reserved)

a field strength of 11.5 T(esla). Until the early 2030s, this upgrade should furnish an increase in the number of collisions and therefore also of events by a factor of 20. The feasibility of future accelerator projects at CERN will among others depend on the results obtained at the LHC. One such project that could start after the completion of the LHC program around 2035 is a linear lepton collider named CLIC. The collisions of electron and positron beams are expected to produce center-of-mass energies of 380–3000 GeV in several steps. A competing e^+e^- project is the ILC (International Linear Collider), probably to be realised in Japan. The initial center-of-mass energy of 250 GeV at the ILC may seem like a step back compared to the LHC energy of 13 TeV but the much smaller rate of "uninteresting" events in e^+e^- collisions makes up for the smaller energy. Both CLIC and ILC should enable precision measurements of the Higgs boson and of the top quark, but they would also search for new particles. Even if the absence of synchrotron radiation in a linear accelerator allows for a relatively cost-effective operation, financing those future projects will remain a major problem. This applies also to the ambitious Chinese project, an enlarged version of the LEP/LHC complex. The CepC (circular electron positron collider, optionally also Chinese electron positron Collider) would have a circular tunnel of 100 km circumference that would in a first step house an e^+e^- storage ring with a total energy of about 240 GeV for precision measurements of the Higgs boson (Higgs Factory). In a possible second step, a proton proton storage ring could follow with a total energy of 50–70 TeV. According to the proponents, e^+e^- collisions could start around 2030.

Intensity Frontier

This area of experimental high energy physics is characterised by high intensities and/or high sensitivities. Neutrino experiments will continue to play an important role here, both with extraterrestrial sources and with neutrinos from reactors and accelerators. An especially ambitious project is the Deep Underground Neutrino Experiment (DUNE). This experiment will use an intensive neutrino beam from the Long-Baseline Neutrino Facility at Fermilab. With a giant underground detector at the Sanford Laboratory in the Homestake gold mine in South Dakota, some 1300 km away from Fermilab, it will also be sensitive to extraterrestrial neutrinos. As in other big neutrino detectors,[2] DUNE also will search for proton decays. Both DUNE and the planned mega-detector Hyper-Kamiokande in Japan with one million tons of water are expected to be operational around 2026. The successful B factories in the U.S. and in Japan with the experiments BaBar and Belle, which confirmed the mechanism of CP violation in the B sector of the Standard Model, will be succeeded by the e^+e^- storage ring Super-KEKB in Tsukuba (Japan). This B factory will deliver a luminosity 40 times higher than at KEKB to the successor experiment Belle II, which started operating in April 2018 and will look in particular for rare decays of B mesons and τ leptons. First results should be available soon. Other high-intensity experiments will be performed with kaons, e.g., in the Japanese facility J-PARC, where also a precision measurement of the anomalous magnetic moment of the muon is being prepared (Chap. 10). Finally, several experiments will try to measure electric dipole moments of particles (electron, neutron) and atoms (mercury, radium, xenon) to obtain additional insight into the mechanism of CP violation. Because the Standard Model predicts in general very small electric dipole moments, which have not been verified experimentally up to now, such experiments are especially promising for finding traces of new physics.

Cosmic Frontier

The search for dark matter and precision measurements of the cosmic microwave background (CMB) are in the focus of astroparticle physics. Although there are strong indirect indications for the existence of dark matter (DM) such as the rotation curves

---

[2]For instance, Kamiokande actually stands for Kamioka n(ucleon)d(ecay)e(xperiment).

of spiral galaxies, gravitational lensing, dynamics of galaxy clusters, etc., all attempts of direct detection have been unsuccessful so far. At present, the second generation of DM experiments is in preparation. Long-time candidates for DM are electrically neutral, massive particles (WIMPs = Weakly Interacting Massive Particles), which only feel gravitational and weak interactions. The goal is to detect their scattering on normal matter. A promising second-generation project is the LZ experiment, a detector with seven tons of liquid xenon that will be housed in the Sanford Laboratory together with the DUNE experiment. In the LZ experiment, scheduled to start in 2020, the recoil of xenon nuclei after scattering with WIMPs will be measured. A smaller experiment called ADMX is dedicated to the hypothetical axion, another DM candidate. In this experiment at the Univ. of Washington (Seattle) axions would be converted into photons after traversing a strong magnetic field. Also in the IceCube Neutrino Observatory in Antarctica the search is on for neutrinos that could originate in the annihilation of WIMPs. A strong indirect evidence for DM comes from CMB measurements that will be continued after the successful satellite experiments COBE, WMAP and PLANCK. The CMB data have furnished additional important results such as the density of the mysterious dark energy, parameters of cosmic inflation and number and masses of neutrinos (e.g., the upper bound (10.3) for the sum of neutrino masses). Presently, already the fourth generation of ground-based CMB experiments called CMB-S4 is in the planning stage. Telescopes in different regions of the earth will be employed. In addition, there are several proposals for CMB experiments with satellites that could be ready from 2025 on.

Quantum Field Theory

The history of the Standard Model is also a success story of renormalisable quantum field theories, which have therefore enjoyed a privileged position in theoretical particle physics for a long time. Supersymmetric extensions of the Standard Model and the many variants of grand unification are all renormalisable quantum field theories. However, by now there are almost countless many of such variants with predictions depending in general on many unknown parameters, making the comparison between theory and experiment more difficult and less significant. Partly for that reason, an alternative approach has gained in importance over the years, where the general structure of extensions of the Standard Model and their phenomenological consequences are analysed without restricting attention to a specific realisation. Grand unification is a good example for this paradigm shift. Since it will probably never be possible to produce particles with masses of the order $10^{16}\,\mathrm{GeV}/c^2$, the actual parameters of a GUT are less interesting than the general recognition that in such theories baryon number can be violated leading in particular to proton decay. To some extent, the situation is similar as at the time of the Fermi theory of the weak interaction before the advent of the Standard Model. Between 1933 and the end of the 1960s, the manifestations of the weak interaction were described in terms of a nonrenormalisable theory, which in the V–A version (Chap. 7) was quite successful although the W boson as carrier of the weak interaction had not been discovered yet. There is however an important difference between the Fermi theory and the Standard Model. Whereas the Fermi theory consisted of a nonrenormalisable quantum field theory only, today we are talking about nonrenormalisable extensions of the renormalisable Standard Model in the framework of an effective quantum field theory (EFT). A special application concerns the scalar sector that is responsible for the spontaneous breaking of the gauge symmetry. Although the simplest realisation of the BEH mechanism with a single scalar doublet (Chap. 9) agrees with all available data, the present experimental accuracy still allows for an extended scalar sector in the TeV range. An EFT by the name of HEFT (Higgs Effective Field Theory) is the proper tool to investigate such extensions in a model independent way.

In a sense, EFTs are the antithesis to the Theory of Everything, the slowly fading long-term goal of string theory. EFTs are the quantum field theoretical realisation of the quantum ladder (Chap. 1). Massive degrees of freedom need not be represented in the theory by explicit quantum fields in order to derive useful predictions at low energies, i.e. at energies substantially smaller than the masses ($\times c^2$) of those very massive particles. For instance, we do not need the still unknown theory of quantum gravity to understand the structure of the hydrogen atom, nor is the precise structure of the electroweak gauge theory relevant for chemistry.

EFTs should be seen as approximations of an underlying "fundamental" theory, which in the sense of the quantum ladder could itself be an EFT of an even more fundamental theory, etc. If the degrees of freedom are known that are relevant at the energy under consideration, the corresponding EFT is usually treated perturbatively. For instance, it does not make much sense to search for an exact solution of the Fermi theory. Also the question of convergence of the perturbative expansion of an EFT is not a relevant issue. The perturbative series must be seen as an asymptotic expansion that ceases to be relevant once explicit contributions of the underlying fundamental theory can no longer be neglected. This recognition also implies that the perturbative expansion of the renormalisable part (here of the Standard Model) must be advanced as much as necessary in order to avoid misinterpreting a discrepancy between theory and experiment as evidence for "New Physics". The domain of applicability of the perturbative series depends most of all on how much bigger the masses ($\times c^2$) in the fundamental theory are than the typical energy scale of the EFT. The great successes of the Standard Model even at LHC energies indicate that in spite of the two-loop approximation (and beyond) of the perturbative series and despite precise experiments we are still not sensitive to the eventual "New Physics". This somewhat sober interpretation of the present situation of particle physics in an EFT setting in no way diminishes the special status of the Standard Model.

As a matter of fact, EFTs appear in many areas of physics, though not always under this label. The effective nature of the theoretical treatment is obvious both in atomic physics and in the physics of condensed matter. But also in particle physics there are interesting applications in addition to the analysis of extensions of the Standard Model. Here, only two such applications will be mentioned, both of which refer to the strong interactions. At energies $<1\,\text{GeV}$, QCD cannot be treated perturbatively because of confinement. The physics of pions and kaons, but also of nucleons at low energies is therefore investigated with an EFT called chiral perturbation theory (Weinberg, Gasser, Leutwyler), where instead of quarks and gluons mesons and baryons are the relevant quantum fields. In general, the symmetries of the underlying fundamental theory are of vital importance for the construction of an EFT, in the concrete case the approximate so-called chiral symmetry of QCD. Chiral perturbation theory has put the physics of hadrons at low energies on a solid basis although it is actually a nonrenormalisable quantum field theory. As strange as it may sound, theorists therefore had to learn how to properly renormalise nonrenormalisable quantum field theories. The often uncontrolled assumptions of the "old" hadronic physics have been replaced in this way by a systematic quantum field theoretical approach.

At LHC energies, the Standard Model can be treated perturbatively. Nevertheless, EFTs play an important role also at those energies. Here the issue is not to replace quarks and gluons by hadron fields. Instead, a major question is how to interpret a multi-particle event as in Fig. 11.2. How can one find indications for physics beyond the Standard Model in such an event? After all, the tracks in the detector reconstructed in Fig. 11.2 do not represent quarks and gluons but charged hadrons and leptons. It is not too surprising that we do not get very far with the standard perturbative approach, not least because of the huge number of particle tracks. Energetic quarks and gluons manifest themselves in the final state of a scattering process as conical hadron bundles called jets whose total momentum points in the direction of the original quark or gluon. No matter how large the energy of the original particle is, in the end this energy spreads over many hadrons with very different energies. This process of hadronisation is therefore a process with very different energy scales that calls for a treatment with EFT methods. The analysis of hadronic jets is an integral part of the interpretation of scattering processes at the LHC.

Fig. 11.2 2-jet event at the LHC with 13 TeV center-of-mass energy (With kind permission of © CERN 2016 for the benefit of the CMS Collaboration. All Rights Reserved)

Some 52 years ago when the author of this book started working on his thesis, the Standard Model was in a way just around the corner and yet it practically played no role in the research landscape of the time. As discussed in earlier chapters, the status of particle physics around 1965 was far from satisfactory. Only ten years later, the Standard Model, one of the great achievements of the human mind, was in full bloom. It has been dominating fundamental physics ever since without interruptions.

It took almost 50 years from the birth of quantum mechanics to the establishment of the Standard Model as known today. This period has brought forth fundamental new insights into the structure of matter and its interactions. From atomic physics with a characteristic resolution of about 10^{-10} m we attained distances of approximately 10^{-17} m in 1970 thanks to the progress of accelerators and of experimental particle physics. Since then, about the same period of time has passed leading to a resolution of at least 10^{-19} m. The comparatively small improvement in resolution by only two orders of magnitude has of course to do with the fact that experimental progress becomes more and more difficult and expensive. That except for the experimental confirmation of the Higgs boson no really fundamental discoveries have been made in recent years has given rise to certain frustrations among particle physicists. In my view, there is no reason to be frustrated. On the one hand, both current and planned activities in experimental particle physics show that the efforts to find hints for "New Physics" have not diminished. On the other hand, the ongoing experimental confirmations of the Standard Model document that with this theory of the fundamental interactions we have reached a level never known before. Although there is no reason for complacency or even vanity, deep satisfaction about the progress achieved is certainly appropriate.

Mathematical Structures, Units and Notation

A

In classical mechanics one investigates the motion of (point-)particles under the in-fluence of certain forces. Newton's second axiom has the form of an equation of motion to determine the time dependence $q(t)$ of the coordinates of the respective particles. Here, $q(t)$ stands for all coordinates. For a single particle three coordinates are needed (e.g., the Cartesian coordinates x, y, z), for N particles $3N$ coordinates. Since we are interested in the temporal development of the coordinates, the New-tonian equation of motion has the form of a differential equation. Thus we have to introduce the derivative of a function. The relevant quantities are well known from everyday life. The velocity $v(t)$ is the first derivative of the coordinate $q(t)$, the acceleration $a(t)$ the second derivative[1]:

$$v(t) = \frac{dq(t)}{dt} \, , \qquad a(t) = \frac{dv(t)}{dt} = \frac{d^2q(t)}{dt^2} \, . \tag{A.1}$$

The revolutionary step of Galilei and Newton was to realise that the acceleration is proportional to the applied force. Hence, the Newtonian equation of motion is a differential equation of second order (m is the mass of the particle, K the force acting on the particle):

$$m\frac{d^2q(t)}{dt^2} = K(q(t), v(t)) \, . \tag{A.2}$$

If your eyelids are beginning to droop here, you should now wake up again. One important manifestation of the unity of physics is that also the basic equations of the fundamental interactions (gravity, electromagnetism, strong and weak nuclear forces) are all differential equations of second order. However, we have to introduce a small generalisation. Already in classical physics, for instance in hydrodynamics and especially in electrodynamics, the quantities of interest are both time and space

---

[1]The exact definition of a derivative can be found in any mathematics textbook for the upper secondary level.

© Springer Nature Switzerland AG 2019

G. Ecker, *Particles, Fields, Quanta*, Undergraduate Lecture Notes in Physics,

https://doi.org/10.1007/978-3-030-14479-1_A

Table A.1 Relations between real-valued quantities a, b

Relation	Interpretation
$a < (>) b$	a smaller (bigger) than b
$a \ll (\gg) b$	a much smaller (bigger) than b
$a \simeq b$	a approximately equal to b
$a \lesssim (\gtrsim) b$	a approximately equal to or smaller (bigger) than b
$a \neq b$	a not equal to b

dependent and are denoted as fields. The electric field $\vec{E}(t, \vec{r})$ and the magnetic field $\vec{B}(t, \vec{r})$ are the best-known examples. In these cases, we speak of field equations instead of equations of motion. Since there are now more than just one independent variable (time t and three space coordinates $\vec{r}$), we have to distinguish between different derivatives. For this purpose one defines partial derivatives as they for instance occur in the Schrödinger equation (3.24):

$$\frac{\partial \psi(t, \vec{r})}{\partial t}, \quad \frac{\partial^2 \psi(t, \vec{r})}{\partial x^2}, \quad \frac{\partial^2 \psi(t, \vec{r})}{\partial y^2}, \quad \frac{\partial^2 \psi(t, \vec{r})}{\partial z^2}. \tag{A.3}$$

Here, $\partial \psi(t, \vec{r})/\partial t$ indicates that the field $\psi(t, \vec{r})$ is differentiated with respect to the time t, keeping the spatial coordinates $\vec{r}$ fixed. All equations of the fundamental interactions are partial differential equations of second order for the respective fields.

The trigonometric functions sine and cosine are assumed to be known. In Euler's formula

$$e^{ix} = \cos x + i \sin x \tag{A.4}$$

they appear as real and imaginary parts of an exponential function. The derivative of the exponential function $de^{ix}/dx = ie^{ix}$ implies the derivatives of the trigonometric functions ($d \sin x/dx = \cos x$, $d \cos x/dx = -\sin x$) and vice versa. Similarly, for the partial derivative of an exponential function of the type e^{ax+b}, where a and b are independent of the variable x, we have $\partial e^{ax+b}/\partial x = ae^{ax+b}$. For relations between real-valued quantities in this book we use the usual symbols collected in Table A.1.

For the description of physical quantities we need a system of units. In almost all industrialised nations the SI system (Système internationale d'unités) must be used for both official and commercial use. The SI system has seven basic units of which we do not need mole and candela. Another one (ampere) we will express in terms of the other basic units. The remaining four basic units are then second, meter, kilogram and kelvin defined as follows.

i. The second (s) is defined as 9192631770 times the period of the transition between the two hyperfine levels of the ground state of atoms of the cesium isotope ^{133}Cs. The period of the transition is the inverse of the transition frequency ν.

ii. The meter (m) is defined as the distance traveled by light during $1/299792458$ of a second in vacuum. Since this definition replaced the original prototype meter, the value of the velocity of light is exact by definition: $c = 299792458$ m/s.

iii. The kilogram (kg) is the mass of the international kilogram prototype kept in Sèvres near Paris. This definition is currently being replaced by another one that will use Planck's constant $\hbar$. Similar to the velocity of light, also $\hbar$ will then have an exact value.

iv. The kelvin (K) as unit of the absolute temperature will only be used in the radiation laws in Chap. 2. Here we content ourselves with an approximate definition that $T = 0$ K corresponds to the more common temperature $-273.15\,°C$ (absolute zero) and that temperature differences measured in K and degree Celsius are identical.

The ampere is the unit of electric current in the SI system. This definition may have its merits for technical applications, but for fundamental physics it is completely unsuitable. For instance, generations of physics students have racked their brains over the mysterious permittivity of the vacuum ϵ_0. Whoever has tried to formulate Lorentz transformations for electric and magnetic fields in the SI system will understand why most theoreticians and practically all particle physicists follow the suggestion of the great Carl Friedrich Gauß to express the electric charge (in the SI system defined by coulomb = ampere-second) in terms of meter, kilogram and second. In the Heaviside system of particle physics, the elementary charge e can most easily be defined in terms of the fine-structure constant α. As a dimensionless quantity, the fine-structure constant has the same value in all systems of units[2]:

$$\alpha = 1/137.035999139(31)\,. \tag{A.5}$$

Here we have used a practical notation especially for quantities measured very precisely where instead of the explicit notation

$$\alpha^{-1} = 137.035999139 \pm 0.000000031 \tag{A.6}$$

one writes the experimental error in brackets as in Eq. (A.5). In the Heaviside system the elementary charge e is then given by

$$e = \sqrt{4\pi\hbar c\alpha}\,. \tag{A.7}$$

With the current value for Planck's constant,[3]

$$\hbar = 1.054571800(13) \cdot 10^{-34}\,\mathrm{J\,s}\,, \tag{A.8}$$

---

[2]Current values of many physical quantities can also be found under http://physics.nist.gov/constants.

[3]Joule is the (derived) unit of energy in the SI system: $\mathrm{J} = \mathrm{kg\,m^2\,s^{-2}}$.

and with the law of error propagation one could now determine the current value of e. We leave this calculation to the gentle reader.

In quantum physics one often uses electron volt (eV) as unit of energy with $1\,\text{eV} = 1.6021766208(98) \cdot 10^{-19}$ J, and the derived quantities $\text{TeV} = 10^3\,\text{GeV} = 10^6\,\text{MeV} = 10^9\,\text{keV} = 10^{12}\,\text{eV}$. Subsequently, nuclear and particle physicists often express also masses in terms of electron volts and the velocity of light. For instance, the electron mass can be written as $m_e = 0.5109989461(31)\,\text{MeV}/c^2$, which appeals more to the imagination of nuclear and particle physicists than the conventional $m_e = 9.10938356(11) \cdot 10^{-31}$ kg.

In atomic and subatomic physics, lengths are often expressed in terms of the following quantities: nm (nanometer), Å (Ångström), fm (femtometer or fermi) with

$$1\,\text{nm} = 10\,\text{Å} = 10^{-9}\,\text{m} = 10^6\,\text{fm} . \tag{A.9}$$

For order-of-magnitude estimates the following relations are useful:

$$GeV^{-1} \simeq 6.58 \cdot 10^{-25}\,\hbar^{-1}\,\text{s} \simeq 1.97 \cdot 10^{-16}\,(\hbar c)^{-1}\,\text{m} \tag{A.10}$$

$$1\,\text{fm} = 10^{-15}\,\text{m} \simeq \frac{\hbar c}{0.2\,GeV} \simeq 3.34 \cdot 10^{-24}\,c\,\text{s} . \tag{A.11}$$

For instance, these relations tell us that light needs about 10^{-24} s to traverse the diameter of a proton (approximately 1 fm). They also imply that an accelerator with energies in the GeV region can resolve distances of about 10^{-16} m. Because the LHC works in the TeV region, particle physics presently enters dimensions smaller than 10^{-19} m.

In phenomenologically oriented particle physics, gravity does not play any role at presently available energies. This can be corroborated with the following dimensional arguments. The strength of gravitational attraction is given by Newton's constant

$$G_\text{N} = 6.67408(31) \cdot 10^{-11}\,\text{m}^3\,\text{kg}^{-1}\,\text{s}^{-2} . \tag{A.12}$$

Except for numerical factors, a unique energy can be constructed from the quantities G_N, $\hbar$, c known as Planck energy:

$$E_\text{P} = \sqrt{\hbar c^5 / G_\text{N}} \simeq 1.22 \cdot 10^{19}\,\text{GeV} . \tag{A.13}$$

The corresponding length

$$l_\text{P} = \sqrt{G_\text{N}\hbar/c^3} \simeq 1.62 \cdot 10^{-35}\,\text{m} \tag{A.14}$$

is known as the Planck length. In our four-dimensional world (one time and three space dimensions), we therefore expect that quantum effects of the still unknown theory of quantum gravity will become relevant only at energies $\gtrsim 10^{19}$ GeV or at distances $\lesssim 10^{-35}$ m. In the foreseeable future, experimental particle physics will stay very far away from these key figures.

Gauge and Lorentz Invariance

B

B.1 Gauge Transformations

Maxwell's equations of classical electrodynamics are a system of partial differential equations for the electric field $\vec{E}(t, \vec{r})$ and the magnetic field $\vec{B}(t, \vec{r})$, depending on the charge density $\rho(t, \vec{r})$ and on the current density $\vec{j}(t, \vec{r})$. In the Heaviside system of particle physics (Appendix A), the equations have the following form:

$$rot\ \vec{E} + \frac{1}{c}\frac{\partial \vec{B}}{\partial t} = 0 \qquad\qquad\qquad div\ \vec{B} = 0 \qquad (\text{B.1a})$$

$$div\ \vec{E} = \rho \qquad\qquad\qquad rot\ \vec{B} - \frac{1}{c}\frac{\partial \vec{E}}{\partial t} = \frac{1}{c}\vec{j} \qquad (\text{B.1b})$$

In cartesian coordinates, the differential operators divergence and curl acting on an arbitrary (differentiable) three-dimensional vector field $\vec{V}(t, \vec{r})$ with components (V_1, V_2, V_3) are defined as follows:

$$div\ \vec{V} = \frac{\partial V_1}{\partial x} + \frac{\partial V_2}{\partial y} + \frac{\partial V_3}{\partial z} \qquad (\text{B.2})$$

$$rot\ \vec{V} = \left(\frac{\partial V_3}{\partial y} - \frac{\partial V_2}{\partial z}, \frac{\partial V_1}{\partial z} - \frac{\partial V_3}{\partial x}, \frac{\partial V_2}{\partial x} - \frac{\partial V_1}{\partial y} \right). \qquad (\text{B.3})$$

The divergence of a vector field is a one-component scalar field, while the curl produces another vector field. Finally, we also need the gradient that turns a scalar field $\Phi(t, \vec{r})$ into a vector field:

$$grad\ \Phi = \left(\frac{\partial \Phi}{\partial x}, \frac{\partial \Phi}{\partial y}, \frac{\partial \Phi}{\partial z} \right). \qquad (\text{B.4})$$

© Springer Nature Switzerland AG 2019
G. Ecker, *Particles, Fields, Quanta*, Undergraduate Lecture Notes in Physics,
https://doi.org/10.1007/978-3-030-14479-1_B

Let us consider first the two field equations in (B.1a), the homogeneous Maxwell equations. Since the divergence of a curl vanishes, there exists a vector field $\vec{A}(t, \vec{r})$ for every (reasonable) magnetic field $\vec{B}(t, \vec{r})$ with

$$\vec{B}(t, \vec{r}) = rot\ \vec{A}(t, \vec{r}) \tag{B.5}$$

so that the Maxwell equation $div\ \vec{B} = 0$ is automatically fulfilled. If we now insert $\vec{B} = rot\ \vec{A}$ into the second homogeneous Maxwell equation, we get

$$rot\left(\vec{E} + \frac{1}{c}\frac{\partial \vec{A}}{\partial t}\right) = 0 . \tag{B.6}$$

The curl of a gradient always vanishes. Thus, under the usual conditions there exists a scalar field that with foresight we call $A_0(t, \vec{r})$ so that

$$\vec{E}(t, \vec{r}) = -grad\ A_0(t, \vec{r}) - \frac{1}{c}\frac{\partial \vec{A}(t, \vec{r})}{\partial t} . \tag{B.7}$$

$A_0(t, \vec{r})$ is called the scalar potential and $\vec{A}(t, \vec{r})$ the vector potential of electrodynamics.

By introducing scalar and vector potentials, we have "solved" the homogeneous Maxwell equations (B.1a) automatically. If we now insert (B.5) and (B.7) into the inhomogeneous Maxwell equations (B.1b), we are left with only four partial differential equations of second order to solve for the potentials $A_0, \vec{A}$. By means of Eqs. (B.5) and (B.7) we can then calculate the physical fields $\vec{B}(t, \vec{r})$ and $\vec{E}(t, \vec{r})$ from the potentials.

Introducing the potentials $A_0, \vec{A}$ simplifies the solution of the Maxwell equations in many cases but the potentials are not uniquely determined. This brings us to gauge transformations that also play an important role in the quantum field theory of the Standard Model. Instead of the original potentials $A_0, \vec{A}$ we define new potentials $A_0', \vec{A}'$ with the help of an arbitrary (differentiable) function $\beta(t, \vec{r})$:

$$A_0' = A_0 + \frac{1}{c}\frac{\partial \beta}{\partial t} , \qquad \vec{A}' = \vec{A} - grad\ \beta . \tag{B.8}$$

If we now calculate the fields $\vec{E}', \vec{B}'$ with the new potentials $A_0', \vec{A}'$ using Eqs. (B.5) and (B.7), we find that

$$\vec{E}'(t, \vec{r}) = \vec{E}(t, \vec{r}) , \qquad\qquad \vec{B}'(t, \vec{r}) = \vec{B}(t, \vec{r}) . \tag{B.9}$$

In other words, the physical fields $\vec{E}, \vec{B}$ remain unchanged under a gauge transformation (B.8). With an appropriate choice of the gauge function $\beta(t, \vec{r})$, solving the Maxwell equations is often facilitated.

The classical Maxwell equations (B.1) are obviously gauge invariant because the potentials do not even appear in the equations. In the Standard Model and in quantum electrodynamics in particular, the situation is different because the photon field, the quantised version of the classical potentials $(A_0, \vec{A})$, appears explicitly in the QED Lagrangian (5.1) and therefore also in the field equations of QED. Using the relativistic notation by combining the scalar and vector potentials into a four-vector field $A(x) = (A_0(t, \vec{r}), \vec{A}(t, \vec{r}))$ with $x = (ct, \vec{r})$, the gauge transformation (B.8) can be written in the (covariant) form

$$A'_\mu(x) = A_\mu(x) + \partial_\mu \beta(x) = A_\mu(x) + \frac{\partial \beta(x)}{\partial x^\mu} \quad (\mu = 0, 1, 2, 3) . \tag{B.10}$$

This is precisely the transformation that guarantees the gauge invariance of the QED Lagrangian (5.1). In the QED Lagrangian the transformation of the photon field is compensated by a phase transformation of the Dirac field. The Lagrangian and the field equations of QED are therefore gauge invariant. As discussed in Chap. 5, gauge invariance of QED guarantees that the (four-component) photon field $A(x)$ only has two physical degrees of freedom as it must be for a massless particle with spin >0 like the photon.

B.2 Lorentz Transformations

Einstein's principle of relativity states that all inertial systems (IS) are equivalent. For definiteness, consider a particle that is at rest in a given IS. In another inertial system IS' it then moves with a constant velocity $\vec{v}$. The Lorentz transformation relates the coordinates of the particle in the original system IS with those in IS'. If we choose our cartesian coordinate system such that the (constant) velocity $\vec{v}$ points in the x-direction, the following relations between the space-time coordinates (t, x, y, z) in IS and (t', x', y', z') in IS' (special Lorentz transformation with velocity $\vec{v} = (v, 0, 0)$) hold:

$$
\begin{aligned}
t' &= \left(t - \frac{vx}{c^2}\right) \Big/ \sqrt{1 - v^2/c^2} \\
x' &= (x - vt) \Big/ \sqrt{1 - v^2/c^2} \\
y' &= y \\
z' &= z .
\end{aligned}
\tag{B.11}
$$

In the limiting case $|v| \ll c$ (all velocities much smaller than the speed of light), the Lorentz transformation (B.11) turns into a Galilei transformation:

$$
\begin{aligned}
t' &= t \\
x' &= x - vt \\
y' &= y \\
z' &= z .
\end{aligned}
\tag{B.12}
$$

In classical mechanics that is invariant with respect to Galilei transformations there exists a universal time ($t' = t$). In contrast, the time in special relativity depends on the IS as can be seen in Eq. (B.11). The dependence of the time on the reference system is the cause of almost all conceptual difficulties of special relativity (Chap. 2).

The Lorentz transformation (B.11) implies the theorem on the addition of velocities in special relativity. Let a particle in IS have a velocity $\vec{w} = (w, 0, 0)$, i.e., it moves in the (positive) x-direction ($w > 0$). Performing a Lorentz transformation with a velocity $\vec{v}$ where $\vec{v} = (-v, 0, 0)$ points for simplicity in the negative x-direction ($v > 0$), the velocity of the particle $\vec{w}' = (w', 0, 0)$ in IS' is given by

$$w' = \frac{w + v}{1 + vw/c^2} \, . \tag{B.13}$$

This equation implies that for a given velocity w the velocity w' increases monotonically with increasing v, but for any $v < c$ we also have $w' < c$. The speed of light in special relativity is a limiting velocity that can never be attained by massive particles.

On the other hand, performing the limit $c \to \infty$ in (B.13) we get the well-known theorem on the addition of velocities in classical mechanics:

$$w' = w + v \, . \tag{B.14}$$

Therefore, the classical addition of velocities is only approximately valid for velocities much smaller than the speed of light.

In a field theory also the fields undergo Lorentz transformations. The simplest case is a scalar field (Table 4.1) that transforms as

$$\varphi'(x') = \varphi(x) \, . \tag{B.15}$$

The transformed scalar field in IS' has the same form in the transformed space-time coordinates as the original field in the original coordinates. For many-component fields such as spinor and vector fields (Table 4.1), the different components transform also among themselves. We will not need the explicit form of those field transformations but they guarantee that a Lagrangian like (5.1) is Lorentz invariant. Therefore, QED and the Standard Model altogether satisfy Einstein's principle of relativity, i.e., they have the same form in all inertial systems.

Index of Scientists

Anderson Carl David (1905–1991)
    American physicist, Nobel Prize 1936, discovery of positron and muon (together with S. H. Neddermeyer)
Anderson Philip Warren (born 1923)
    American physicist, Nobel Prize 1977, spontaneous symmetry breaking, super-conductivity, antiferromagnetism
Bahcall John Norris (1934–2005)
    American astrophysicist, calculations of the solar neutrino flux (Standard Solar Model SSM), initiator of the Hubble space telescope
Balmer Johann Jakob (1825–1898)
    Swiss mathematician and physicist, Balmer series
Becquerel Antoine-Henri (1852–1908)
    French physicist, Nobel Prize 1903, discovery of radioactivity, spectroscopy
Bell John Stewart (1928–1990)
    Irish physicist, CPT theorem, Adler-Bell-Jackiw anomaly, Bell inequality, foundations of quantum mechanics
Bethe Hans Albrecht (1906–2005)
    American physicist of German origin, Nobel Prize 1967, Bethe-Weizsäcker formula (droplet model of atomic nuclei), Bethe-Weizsäcker cycle, Bethe-Heitler formula, Lamb shift, Bethe–Salpeter equation, Brueckner-Bethe theory
Bjorken James Daniel (born 1934)
    American physicist, prediction of the charm quark, Bjorken scaling behaviour, parton model, textbook on quantum field theory (with S. Drell)
Bohr Niels Hendrik David (1885–1962)
    Danish physicist and philosopher, Nobel Prize 1922, atomic model, Copenhagen interpretation of quantum mechanics, Bohr magneton
Born Max (1882–1970)
    German physicist and mathematician, Nobel Prize 1954, co-founder of quantum mechanics, statistical interpretation of the wave function, Born–Oppenheimer approximation, Born approximation

© Springer Nature Switzerland AG 2019

G. Ecker, *Particles, Fields, Quanta*, Undergraduate Lecture Notes in Physics,
https://doi.org/10.1007/978-3-030-14479-1_C

Bose Satyendranath (1894–1947)
  Indian physicist, Bose–Einstein statistics, boson, Bose–Einstein condensation
Bragg William Henry (1862–1942) and William Lawrence (1890–1971)
  British physicists (father and son), Nobel Prize 1915, Bragg equation, Bragg spectrometer
Broglie Louis de (1892–1987)
  French physicist, Nobel Prize 1929, matter waves
Brout Robert (1928–2011)
  Belgian physicist of American origin, cosmic inflation, Brout-Englert-Higgs mechanism
Brown Robert (1773–1858)
  Scottish botanist, Brownian motion
Cabibbo Nicola (1935–2010)
  Italian physicist, physics of electron-positron collider, Cabibbo angle, Cabibbo–Kobayashi–Maskawa mixing matrix (CKM matrix), lattice gauge theories
Cavendish Henry (1731–1810)
  British scientist, discovery of hydrogen, determination of the gravitational constant and of the earth mass
Chadwick James (1891–1974)
  British physicist, Nobel Prize 1935, continuous electron spectrum in β decay, first indications for strong nuclear force, discovery of the neutron
Chew Geoffrey Foucar (born 1924)
  American physicist, proponent of S-Matrix theory (bootstrap method), Chew–Frautschi plot
Coleman Sidney Richard (1937–2007)
  American physicist, Coleman theorem, Coleman-Weinberg potential, Coleman–Mandula theorem, Lectures on particle physics (Erice Summer Schools)
Compton Arthur Holly (1892–1962)
  American physicist, Nobel Prize 1927, Compton effect, Compton wave length
Cooper Leon Neil (born 1930)
  American physicist, Nobel Prize 1972, Cooper pairs, BCS theory of superconductivity
Coulomb Charles Augustin de (1736–1806)
  French physicist, electro- and magnetostatics, Coulomb potential
Cowan Clyde Lorrain (1919–1974)
  American physicist, detection of electron neutrinos (with F. Reines)
Cronin James Watson (1931–2016)
  American physicist, Nobel Prize 1980, co-discoverer of CP violation in decays of neutral kaons, cosmic radiation, Cronin effect
Curie Marie Sklodowska (1867–1934)
  French physicist and chemist of Polish origin, Nobel Prize for physics 1903 and for chemistry 1911, discovered the elements polonium and radium together with her husband Pierre Curie (1859–1906, French physicist, Nobel Prize for physics 1903)

Dalitz Richard Henry (1925–2006)
  Australian physicist, π^0 decay (Dalitz pair), Dalitz plot, quark model
Davis Raymond Jr. (1914–2006)
  American chemist and physicist, Nobel Prize 2002 for physics, first experiment
  for detection of solar neutrinos
Davisson Clinton Joseph (1881–1958)
  American physicist, Nobel Prize 1937, diffraction of matter waves (together with
  L. H. Germer)
Dirac Paul Adrien Maurice (1902–1984)
  British physicist, Nobel Prize 1933, co-founder of quantum theory, Dirac equation,
  Fermi-Dirac statistics, antimatter, quantum field theory
Drell Sidney David (1926–2016)
  American physicist, Drell-Yan process, textbook on quantum field theory (with
  J. D. Bjorken)
Dyson Freeman John (born 1923)
  British–American physicist and mathematician, quantum electrodynamics, renor-
  malisation theory, spin waves, Dyson sphere
Ehrenfest Paul (1880–1933)
  Austrian physicist, statistical mechanics, Ehrenfest theorem, Ehrenfest paradox
Einstein Albert (1879–1955)
  physicist of German origin, Nobel Prize 1921, the scientific genius per se, special
  and general theories of relativity, photo effect, Brownian motion, Bose–Einstein
  statistics
Ellis Charles Drummond (1895–1980)
  British physicist, investigation of nuclear β decays
Englert François Baron (born 1932)
  Belgian physicist, Nobel Prize 2013, cosmic inflation, Brout-Englert-Higgs mech-
  anism
Euler Leonhard (1707–1783)
  Swiss mathematician and physicist, calculus (of variations), number theory, Euler
  number, hydrodynamics, optics
Everett Hugh III (1930–1982)
  American physicist, many-worlds interpretation of quantum theory
Faraday Michael (1791–1867)
  British physicist and chemist, electromagnetic induction, diamagnetism, laws of
  electrolysis
Fermi Enrico (1901–1954)
  Italian physicist, Nobel Prize 1938, nuclear physics, Fermi–Dirac statistics, Fermi
  surface, Fermi gas, Golden Rule, Thomas–Fermi theory, theory of β decay (Fermi
  interaction), first nuclear reactor
Feynman Richard Phillips (1918–1988)
  American physicist, Nobel Prize 1965, quantum electrodynamics, Feynman dia-
  grams, path integral, V–A theory, parton model, proposal of a quantum computer,
  Feynman Lectures of Physics

Fierz Markus (1912–2006)
   Swiss physicist, spin-statistics theorem, quantum field theory
Fitch Val Logsdon (1923–2015)
   American physicist, Nobel Prize 1980, co-discoverer of CP violation in decays
   of neutral kaons
Fourier Jean Baptiste Joseph (1768–1830)
   French mathematician and physicist, Fourier analysis, thermal conduction
Franck James (1882–1964)
   American physicist of German origin, Nobel Prize 1925, Franck–Hertz experi-
   ment, Franck–Condon principle
Friedman Jerome Isaac (born 1930)
   American physicist, Nobel Prize 1990, deep inelastic electron–nucleon scattering
   (SLAC-MIT experiment)
Fritzsch Harald (born 1943)
   German physicist, quantum chromodynamics, grand unification
Furry Wendell Hinkle (1907–1984)
   American physicist, Furry theorem
Galilei Galileo (1564–1642)
   Italian physicist, mathematician and astronomer, laws of falling bodies, Jupiter
   moons, phases of Venus, one of the founders of modern physics: experiment as a
   question to nature
Gasser Jürg (born 1944)
   Swiss physicist, chiral perturbation theory, determination of quark masses
Gauß Carl Friedrich (1777–1855)
   German mathematician, astronomer and physicist, Princeps Mathematicorum,
   non-Euclidean geometry, number theory, theory of errors, normal distribution,
   measurement of earth magnetic field
Geiger Johannes Wilhelm (1882–1945)
   German physicist, Geiger–Marsden experiments, Geiger counter
Gell-Mann Murray (born 1929)
   American physicist, Nobel Prize 1969, strangeness, V–A theory, Gell-Mann–Low
   equation, dispersion relations, chiral symmetry, classification of hadrons (eight-
   fold way), quark model, current algebra, quantum chromodynamics
Gerlach Walther (1889–1979)
   German physicist, nuclear physics, Stern–Gerlach experiment
Glashow Sheldon Lee (born 1932)
   American physicist, Nobel Prize 1979, electroweak gauge theory, prediction of
   the charm quark, GIM mechanism, charmonium, grand unification
Goldberger Marvin Leonard (1922–2014)
   American physicist, scattering theory, dispersion relations, Goldberger-Treiman
   relation
Goldstone Jeffrey (born 1933)
   British physicist, Goldstone theorem, Nambu–Goldstone bosons, string theory,
   solitons

Gordon Walter (1893–1939)
  German physicist, Klein–Gordon equation
Goudsmit Samuel Abraham (1902–1978)
  American physicist of Dutch origin, electron spin
Gross David Jonathan (born 1941)
  American physicist, Nobel Prize 2004, Gross-Llewellyn-Smith sum rule, asymptotic freedom of Yang–Mills theories, quantum chromodynamics, instantons and confinement, Gross–Neveu model, string theory
Hamilton William Rowan (1805–1865)
  Irish physicist and mathematician, geometrical optics, Hamiltonian mechanics
Heaviside Oliver (1850–1925)
  British mathematician and physicist, Heaviside system of units, Heaviside step function, Kennelly–Heaviside layer of the ionosphere
Heisenberg Werner Karl (1901–1976)
  German physicist, Nobel Prize 1932, co-founder of quantum mechanics (matrix mechanics), uncertainty relation, nuclear physics (isospin), S-matrix, ferromagnetism
Hertz Heinrich (1857–1897)
  German physicist, experimental verification of electromagnetic waves
Hertz Gustav Ludwig (1887–1975)
  German physicist, Nobel Prize 1925, Franck–Hertz experiment
Higgs Peter Ware (born 1929)
  British physicist, Nobel Prize 2013, Brout-Englert-Higgs mechanism, Higgs boson
't Hooft Gerardus (born 1946)
  Dutch physicist, Nobel Prize 1999, renormalisation of non-abelian gauge theories, quark confinement, anomalies of quantum field theory, instantons, monopoles in Yang–Mills theories, holographic principle
Hund Friedrich Hermann (1896–1997)
  German physicist, Hund's rules, tunnel effect, nuclear spectra
Iliopoulos John (born 1940)
  Greek physicist, GIM mechanism, anomaly freedom of electroweak gauge theory (with C. Bouchiat and P. Meyer)
Jeans James Hopwood (1877–1946)
  British physicist, mathematician and astronomer, Rayleigh–Jeans law, astrophysics, cosmology
Jolly Philipp von (1809–1884)
  German physicist, precision experiment for the law of gravity
Jordan Ernst Pascual (1902–1980)
  German physicist, matrix mechanics, transformation theory, quantum field theory, cosmology
Kajita Takaaki (born 1959)
  Japanese physicist, Nobel Prize 2015, discovery of neutrino oscillations in atmospheric neutrinos (Super-Kamiokande experiment)

Lewis Gilbert Newton (1875–1946)
   American physical chemist, light quantum renamed photon, chemical binding, thermodynamics
Llewellyn Smith Christopher (born 1942)
   British physicist, electroweak gauge theory, quark-gluon parton model, Gross-Llewellyn-Smith sum rule
Lorentz Hendrik Antoon (1853–1928)
   Dutch physicist and mathematician, Nobel Prize 1902, electron theory of matter, Lorentz force, Lorentz transformation
Lüders Gerhart (1920–1995)
   German physicist, CPT theorem, spin-statistics theorem
Mach Ernst Waldfried Josef Wenzel (1838–1916)
   Austrian physicist, psychologist and philosopher, co-founder of empiriocriticism, Mach cone, Mach number
Maiani Luciano (born 1941)
   Italian physicist, GIM mechanism, nonleptonic decays, lattice QCD
Majorana Ettore (1906–1938?)
   Italian physicist, nuclear physics, theory of neutrinos, Majorana neutrino, Majoron
Marshak Robert Eugene (1916–1992)
   American physicist, V–A theory, astrophysics, nuclear physics
Maskawa Toshihide (born 1940)
   Japanese physicist, Nobel Prize 2008, CP violation, Cabibbo–Kobayashi–Maskawa mixing matrix (CKM matrix)
Maxwell James Clerk (1831–1879)
   Scottish physicist, fundamental contributions to electrodynamics and statistical mechanics, Maxwell equations, Maxwell–Boltzmann distribution
McDonald Arthur Bruce (born 1943)
   Canadian physicist, Nobel Prize 2015, discovery of neutrino oscillations in solar neutrinos (Sudbury Neutrino Observatory)
Meißner Walther (1882–1974)
   German physicist, Meißner–Ochsenfeld effect
Michelson Albert Abraham (1852–1931)
   American physicist, Nobel Prize 1907, Michelson interferometer, Michelson-Morley experiment (measurement of the speed of light)
Minkowski Peter (born 1941)
   Swiss physicist, grand unification, SO(10), seesaw mechanism for neutrino masses
Nambu Yoichiro (1921–2015)
   American physicist of Japanese origin, Nobel Prize 2008, spontaneous symmetry breaking, chiral symmetry of strong interaction, colour degree of freedom, string theory
Newton Isaac (1642–1726)
   British physicist and mathematician, one of the most influential scientists of all times, universal law of gravitation, axioms of classical mechanics, calculus
Nishina Yoshio (1890–1951)
   Japanese physicist, Klein–Nishina formula

Noether Emmy (1882–1935)
  German mathematician, Noether theorem, algebraic topology
Ochsenfeld Robert (1901–1993)
  German physicist, ferromagnetism, Meißner–Ochsenfeld effect
Oppenheimer Julius Robert (1904–1967)
  American physicist, Born–Oppenheimer approximation, neutron stars, gravitational collapse (black holes), Manhattan project
Pauli Wolfgang Ernst (1900–1958)
  physicist of Austrian origin, Nobel Prize 1945, exclusion principle, prediction of the neutrino, quantum field theory, spin-statistics theorem, CPT theorem
Perl Martin Lewis (1927–2014)
  American physicist, Nobel Prize 1995, discovery of the τ lepton
Perlmutter Saul (born 1959)
  American astrophysicist, Nobel Prize 2011, accelerated expansion of the universe, supernovas
Perrin Jean-Baptiste (1870–1942)
  French physicist, Nobel Prize 1926, experiments on Brownian motion, determination of Avogadro's constant
Planck Max Karl Ernst Ludwig (1858–1947)
  German physicist, Nobel Prize 1918, founder of quantum physics, Planck's radiation law, Planck's constant
Poincaré Jules Henri (1854–1912)
  French mathematician and physicist, algebraic topology, complex analysis, celestial mechanics, Poincaré group
Politzer Hugh David (born 1949)
  American physicist, Nobel Prize 2004, asymptotic freedom of Yang–Mills theories, quantum chromodynamics, charmonium
Pomeranchuk Isaak Jakowlewitsch (1913–1966)
  Russian physicist, cosmic rays, reactor physics, synchrotron radiation, superfluidity, Pomeranchuk theorem, S-matrix theory, Pomeron
Powell Cecil Frank (1903–1969)
  British physicist, Nobel Prize 1950, discovery of the charged pion (with G. Occhialini, H. Muirhead and C. Lattes)
Prout William (1785–1850)
  British physician and chemist, Prout's rule
Rabi Isidor Isaac (1898–1988)
  American physicist, Nobel Prize 1944, resonance method for investigation of magnetic properties of nuclei, Rabi oscillations
Reines Frederick (1918–1998)
  American physicist, Nobel Prize 1995, detection of electron neutrinos (with C. Cowan), cosmic neutrinos, double β decay, supernova neutrinos
Richter Burton (1931–2018)
  American physicist, Nobel Prize 1976, discovery of J/ψ meson ($c\bar{c}$ bound state), electron–positron collider SPEAR (SLAC)

Riess Adam Guy (born 1969)
    American astrophysicist, Nobel Prize 2011, accelerated expansion of the universe, supernovas
Rubbia Carlo (born 1934)
    Italian physicist, Nobel Prize 1984, discovery of W and Z bosons (UA1 experiment CERN)
Rutherford Ernest (1871–1937)
    British physicist born in New Zealand, Nobel Prize for chemistry 1907, Rutherford model of the atom, artificial radioactivity, discovery of the proton, prediction of the neutron
Rydberg Johannes (1854–1919)
    Swedish physicist, Rydberg constant, Rydberg atoms
Salam Abdus (1926–1996)
    Pakistani physicist, Nobel Prize 1979, electroweak gauge theory, renormalisation theory, spontaneous symmetry breaking, Pati-Salam model (GUT), supersymmetry (superspace)
Schmidt Brian Paul (born 1967)
    American astrophysicist, Nobel Prize 2011, accelerated expansion of the universe, supernovas
Schrödinger Erwin Rudolf Josef Alexander (1887–1961)
    Austrian physicist, Nobel Prize 1933, co-founder of quantum mechanics (wave mechanics), Schrödinger equation, Schrödinger's cat
Schwartz Melvin (1932–2006)
    American physicist, Nobel Prize 1988, discovery of the muon neutrino (with L. Lederman and J. Steinberger), CP violation
Schwinger Julian Seymour (1918–1994)
    American physicist, Nobel Prize 1965, quantum electrodynamics, Lippmann-Schwinger equation, Lamb shift, anomalous magnetic moment, Dyson-Schwinger equation, Rarita- Schwinger equation, Schwinger model
Soddy Frederick (1877–1956)
    British chemist, Nobel Prize 1921, isotopes
Sommerfeld Arnold Johannes Wilhelm (1868–1951)
    German mathematician and physicist, fine structure of spectral lines, fine-structure constant, Bohr-Sommerfeld model
Steinberger Jack (Hans Jakob) (born 1921)
    American physicist of German origin, Nobel Prize 1988, π^0 lifetime, Σ hyperon, discovery of the muon neutrino (with L. Lederman and M. Schwartz), detection of direct CP violation (K^0 decays), neutrino scattering
Stern Otto (1888–1969)
    American physicist of German origin, Nobel Prize 1943, Stern–Gerlach experiment, molecular beam method, magnetic moment of the proton
Strutt John William (Lord Rayleigh, 1842–1919)
    British physicist, Nobel Prize 1904, discovery of the element argon, Rayleigh–Jeans law, Rayleigh scattering

Sudarshan Ennackal Chandy George (1931–2018)
   Indian physicist, V–A theory, quantum optics
Taylor Richard Edward (1929–2018)
   Canadian physicist, Nobel Prize 1990, deep inelastic electron–nucleon scattering
   (SLAC-MIT experiment)
Telegdi Valentine Louis (1922–2006)
   American physicist of Hungarian origin, verification of parity violation,
   Bargmann–Michel–Telegdi equation, muonic atoms
Thirring Walter (1927–2014)
   Austrian physicist, Thirring model, dispersion theory, stability of matter
Thomson William (Lord Kelvin, 1824–1907)
   British physicist, thermodynamics, Joule–Thomson effect
Thomson Joseph John (1856–1940)
   British physicist, Nobel Prize 1906, discovery of the electron, Thomson model of
   the atom (plum pudding model)
Thomson George Paget (1892–1975)
   British physicist (son of J. J. Thomson), Nobel Prize 1937, experiment for wave
   nature of the electron (with A. Reid), nuclear physics
Ting Samuel Chao Chung (born 1936)
   American physicist, Nobel Prize 1976, discovery of the J/ψ meson ($c\bar{c}$ bound
   state), Alpha-Magnet-Spectrometer (cosmic rays)
Tomonaga Shinichiro (1906–1979)
   Japanese physicist, Nobel Prize 1965, quantum electrodynamics, renormalisation
   theory, nuclear physics, Tomonaga–Luttinger model
Uhlenbeck George Eugene (1900–1988)
   American physicist of Dutch origin, electron spin
van der Meer Simon (1925–2011)
   Dutch physicist, Nobel Prize 1984, stochastic cooling
Veltman Martinus Justinus Godefriedus (born 1931)
   Dutch physicist, Nobel Prize 1999, renormalisation of non-abelian gauge theories
Weinberg Steven (born 1933)
   American physicist, Nobel Prize 1979, electroweak gauge theory, current algebra,
   chiral symmetry, Standard Model of fundamental interactions
Weiss Pierre-Ernest (1865–1940)
   French physicist, para- and ferromagnetism, Weiss domains, Curie–Weiss law
Weisskopf Victor Frederick (1908–2002)
   American physicist of Austrian origin, quantum mechanics, quantum electrody-
   namics, Lamb shift, theoretical nuclear physics, MIT bag model
Weizsäcker Carl Friedrich von (1912–2007)
   German physicist and philosopher, Bethe–Weizsäcker formula (droplet model of
   atomic nuclei), Bethe–Weizsäcker cycle, philosophical aspects of quantum theory
Wess Julius (1934–2007)
   Austrian physicist, $SU(3)$ classification of hadrons, supersymmetry, Wess–
   Zumino model, conformal symmetry, Wess–Zumino–Witten model

Wheeler John Archibald (1911–2008)
  American physicist, nuclear physics, S-matrix, general relativity (inventor of the
  name "black hole"), Wheeler–DeWitt equation
Wien Wilhelm (1864–1928)
  German physicist, Nobel Prize 1911, Wien's displacement law
Wigner Eugene Paul (1902–1995)
  American physicist of Hungarian origin, Nobel Prize 1963, group theory and
  quantum mechanics, quantum field theory, parity, nuclear physics, Breit–Wigner
  formula, Wigner-Eckart theorem
Wilczek Frank Anthony (born 1951)
  American physicist, Nobel Prize 2004, asymptotic freedom of Yang–Mills theo-
  ries, quantum chromodynamics, axion, quark matter, cosmology
Wu Chien-Shiung (1912–1997)
  American physicist of Chinese origin, verification of parity violation
Yang Chen Ning (born 1922)
  American physicist of Chinese origin, Nobel Prize 1957, statistical mechanics
  (Ising model), Yang–Mills theory, parity violation
Yukawa Hideki (1907–1981)
  Japanese physicist, Nobel Prize 1949, prediction of mesons and of an intermediate
  vector boson, Yukawa potential
Zeeman Pieter (1865–1943)
  Dutch physicist, Nobel Prize 1902, magnetism and radiation, Zeeman effect
Zumino Bruno (1923–2014)
  Italian physicist, CPT theorem, spin-statistics theorem, supersymmetry, Wess-
  Zumino model, Wess-Zumino-Witten model
Zweig George (born 1937)
  American physicist, quark model (independently of Gell-Mann), Zweig rule

Glossary

Action   Quantity measured in units $J s = kg\, m^2\, s^{-1}$. Energy × time, space coordinate × momentum, angular momentum all have the dimension of an action. In quantum theory, Planck's constant $\hbar$ is the fundamental unit of the action. In (quantum) field theory, the four-dimensional integral of the → Lagrangian is also called action. Not surprisingly, the action has the dimension of an action

Antimatter   The CPT theorem (Chap. 4) states that in relativistic quantum field theories every particle has an associated antiparticle with the same mass and opposite electric charge. Electrically neutral particles may be their own antiparticles as for instance the photon

Boson   Particle or bound state with integer spin (in units of $\hbar$); bosonic many-particle states satisfy Bose–Einstein statistics (Chap. 4)

Causality   Cause (time t_c) ⇒ effect (time t_e) with $t_c < t_e$ in nonrelativistic physics. In special relativity time depends on the inertial system: causality amounts to the condition that signals cannot be transmitted with superluminal velocities

Confinement   Quarks and gluons are permanently confined in → hadrons. In addition to experimental evidence, many theoretical arguments support confinement but a direct proof using only the QCD field equations is still missing

Divergence   Infinities occurring in the perturbative evaluation of amplitudes (→ S-matrix elements), caused by the unknown structure of physics at smallest distances (highest energies). The → renormalisation program of quantum field theory shifts this unknown structure to masses and coupling constants, which must be determined experimentally

Electroweak theory   Unified → gauge theory of electromagnetic and weak interactions (Chap. 7); together with → quantum chromodynamics, this theory constitutes the Standard Model of fundamental interactions

Exclusion principle   Consequence of the spin-statistics theorem (Chap. 4) of relativistic quantum field theories: two identical → fermions cannot be in the same quantum → state. Originally postulated for electrons by Pauli to understand atomic structure

© Springer Nature Switzerland AG 2019

G. Ecker, *Particles, Fields, Quanta*, Undergraduate Lecture Notes in Physics,
https://doi.org/10.1007/978-3-030-14479-1

Fermion   Particle or bound state with half-integer spin (in units of $\hbar$); fermionic many-particle states satisfy Fermi–Dirac statistics (Chap. 4)

Gauge group   Set of local $\rightarrow$ symmetry transformations

Gauge invariance $\rightarrow$ Appendix B

Gauge theory   (Quantum) field theory that is invariant with respect to local symmetry transformations ($\rightarrow$ gauge group). Gauge invariance (Appendix B) of a relativistic quantum field theory requires the existence of gauge bosons with $\rightarrow$ spin 1, in the Standard Model photon, gluons, W and Z bosons

Hadron   Hadrons are particles or bound states affected by the strong interactions. One distinguishes between mesons (integer spin) and baryons (half-integer spin). All hadrons are bound states of quarks and gluons (Chaps. 8, 9)

Interaction   The nonrelativistic concept of force is replaced in the relativistic domain by the more comprehensive concept of interaction. According to modern physics, all physical phenomena can be related to exactly four fundamental interactions: gravitation, electromagnetism and the strong and weak nuclear interactions

Lagrangian   Generalisation of the Lagrange function of classical mechanics in field theory. Compact representation of a (quantum) field theory from which the field equations can be derived

Lepton   Leptons are particles that are not affected by the strong interactions. There are three types of charged leptons and their associated neutrinos: electron, muon, tau lepton (Chap. 9)

Lorentz invariance $\rightarrow$ Appendix B

Perturbation theory   Expansion of an amplitude ($\rightarrow$ S-matrix element) in quantum field theory in powers of one or several coupling constants, e.g., in powers of the fine-structure constant α in QED

Quantum chromodynamics (QCD)   Quantum field theory of the strong interactions (strong nuclear force, Chap. 8), $\rightarrow$ gauge theory with eight gluons as gauge bosons

Quantum electrodynamics (QED)   Quantum field theory of the electromagnetic interaction (Chap. 5), $\rightarrow$ gauge theory with the photon as gauge boson

Renormalisation   Parameters of a quantum field theory (masses, coupling constants) must be related to measurable quantities; those relations are in general modified (renormalised) by the interaction(s). Renormalisation removes the $\rightarrow$ divergences of S-matrix elements occurring in $\rightarrow$ perturbation theory (Chap. 6). Renormalisable quantum field theories give rise to well-defined predictions at each order in perturbation theory

S-matrix   Unitary (infinite-dimensional) matrix; matrix elements are the probability amplitudes for the transitions from given initial states ($t \rightarrow -\infty$) to specified final states ($t \rightarrow \infty$). Absolute squares of S-matrix elements determine measurable quantities like cross sections and decay probabilities

Spin   Intrinsic angular momentum with the physical dimension of an $\rightarrow$ action, conventionally given in units of $\hbar$. Unlike the orbital angular momentum that can adopt only integer values, the spin can also have half-integer values (representation theory of the rotation group)

State   Description of a quantum system in terms of kinematical quantities (energy, momentum) and quantum numbers (charge, spin, …). In quantum mechanics, the Schrödinger equation determines the time development of a quantum state. In relativistic quantum field theories, the temporal development is much more complex (particle creation and annihilation). Experimentally accessible are probabilities for transitions from given initial states to specified final states ($\rightarrow$ S-matrix)

Symmetry   Transformation of coordinates and fields leaving equations of motion and/or field equations invariant. Such transformations satisfy the postulates of a mathematical group: the result of two successive symmetry transformations is again a symmetry transformation. Depending on the transformation parameters, one distinguishes between discrete (e.g., space reflection) and continuous (e.g., rotations) symmetries. Those parameters are either coordinate independent (global symmetry) or they depend on local coordinates (local symmetry = gauge symmetry)

Index

A

Action, 11

Antimatter, 13, 29, 34, 38, 41, 43, 45, 49, 55, 60, 69, 89, 93, 97, 100–101, 110

Asymptotic expansion, 54, 113

Asymptotic freedom, 81–84, 87

ATLAS/CMS experiments, 94, 102, 109

Atomic model
  Bohr, 4, 8, 17–20, 23, 30, 66
  Bohr–Sommerfeld, 20, 28
  Rutherford, 14–15, 17, 66
  Thomson, 14, 66

Axion, 112

B

BEH mechanism, 74, 91–93, 101, 105, 113

Beta decay, 65–71

B meson, 111

Bohr radius, 18–19

Bootstrap, 79

Born approximation, 48–49, 55, 58, 60–61, 70, 89, 92

Bose-Einstein condensation, 37, 74

Bottom quark, 91

Branching ratio, 2

Brownian motion, 11

C

Causality, 33–36

CepC, 110

CERN, 2, 42, 58, 74, 88, 92, 94, 110

Charge conjugation, 41, 45, 69

Charge screening, 80–84

Charm quark, 89

Chiral perturbation theory, 114

CKM matrix, 100–101

Classical electron radius, 18, 50

CLIC, 110

Colour, 82–84, 90, 91, 106

Commutation relations, 23–27, 35–38

Compton wave length, 18, 71, 74

Confinement, 85, 93, 104, 113

Conservation law, 21, 39–42

Copenhagen interpretation, 1, 26

Correspondence principle, 22, 23, 41, 43, 46

Cosmic frontier, 109–112

Cosmic Microwave Background (CMB), 102, 111

CP symmetry, 42, 70, 91, 101, 111

CPT theorem, 41, 42, 69, 90

Crossing symmetry, 49, 55, 71

Cross section, 50, 71, 95, 97, 104
  Compton, 50–51, 60
  deep inelastic, 80
  Thomson, 50, 60

D

Dark energy, 102, 112

Dark matter, 87, 102, 109, 111

De Broglie wave length, 21

Deep Underground Neutrino Experiment (DUNE), 111–112

Degeneracy, 40

Dirac equation, 29–30, 44, 52, 57, 97

© Springer Nature Switzerland AG 2019

G. Ecker, *Particles, Fields, Quanta*, Undergraduate Lecture Notes in Physics,
https://doi.org/10.1007/978-3-030-14479-1

Printed in the United States
By Bookmasters